Soil Mechanics Laboratory Manual

Fourth Edition

Braja M. Das
Dean
School of Engineering
California State University, Sacramento

Engineering Press

San Jose, California

Printed in United States of America   4 3

Library of Congress Cataloging-in-Publication Data

Das, Braja M., 1941–
      Soil mechanics laboratory manual / Braja M. Das. —4th ed.
         p.   cm.
      Includes bibliographical references.
      ISBN 0-910554-87-0
      1. Soil mechanics—Laboratory manuals. I. Title.
TA710.5.D28   1992
624.1'5136'078—dc20   92-12291
                                                                                    CIP

Engineering Press   **P.O. Box 1**   **San Jose, California 95103-0001**

Preface

Since the early 1940's the study of soil mechanics has made great progress all over the world. A course in soil mechanics is presently required for all undergraduate students in four-year civil engineering or civil engineering technology programs in the United States. It usually includes some laboratory procedures that are essential in understanding the properties of soils and their behavior under stress and strain: the present laboratory manual is prepared for classroom use by undergraduate students taking such a course.

The procedures and equipment described in this manual are fairly common. For a few tests, such as permeability, direct shear, and unconfined compression, the existing equipment in a given laboratory may differ slightly. In those cases, it is necessary that the instructor familiarize students with the operation of the equipment. Triaxial test assemblies are costly, and the equipment varies widely. For that reason, only general outlines for triaxial tests are presented.

For each laboratory test procedure described, sample calculation(s) and graph(s) are included. Also, blank tables for each test are provided at the end of the manual for student use.

Appendix A presents the soil classification systems. The Unified Soil Classification System has been updated to conform to the most recent ASTM specifications.

In some instances, instructors require that computers be used for calculation, of the final results, based on laboratory observations. For this reason, two disks have been included in this edition. This diskette has been prepared by my colleague, Aslam Kassimali, Associate Professor in the Department of Civil Engineering and Mechanics, Southern Illinois University at Carbondale. Instructions for using this diskette are given in Appendix B. This software will enable students to generate completed tables and some graphs in the same format as presented in the text. It is important to note that instructors and students can choose to use either hand calculators and tables provided at the end of the manual or the computer software to analyze the data. Even if students use the computer software, some hand calculator use is essential in order to give students a feel for data reduction.

During the preparation of the first edition of the manual, Professors Paul C. Hassler and the late Andrew D. Jones of the University of Texas at El Paso provided continuous encouragement. I would also like to thank Dr. Haskell Monroe, University of Missouri-Columbia, for his support, understanding, and encouragement in the development of my professional career during the past fifteen years. Last but not least is my wife, Janice F. Das, who apparently possesses endless energy and enthusiasm. Not only did she type the manuscript, she also prepared most of the tables, graphs, and other line drawings.

Braja M. Das

To Janice and Valerie

Contents

1. Laboratory Test and Report Preparation   **1**

2. Determination of Moisture Content   **5**

3. Specific Gravity   **9**

4. Sieve Analysis   **15**

5. Hydrometer Analysis   **23**

6. Liquid Limit Test   **35**

7. Plastic Limit Test   **43**

8. Shrinkage Limit Test   **47**

9. Constant Head Permeability Test in Sand   **53**

10. Falling Head Permeability Test in Sand   **59**

11. Standard Proctor Compaction Test   **63**

12. Modified Proctor Compaction Test   **71**

13. Determination of Field Unit Weight of Compaction by Sand Cone Method   **75**

14. Direct Shear Test on Sand   **81**

15. Unconfined Compression Test   **89**

16. Consolidation Test   **97**

17. Triaxial Tests in Clay   **109**

References   **125**

Appendices

A   Engineering Classification of Soils   **127**

B   Computer Software   **145**

C   Tables for Laboratory Work   **167**

1
Laboratory Test and Report Preparation

Proper laboratory testing of soils for determination of their physical properties is an integral part in the design and construction of structural foundations, the placement and improvement of soil properties, and the specification and quality control of soil compaction works. It needs to be kept in mind that natural soil deposits often exhibit a high degree of nonhomogenity. The physical properties of a soil deposit can change to a great extent even within a few hundred feet. The fundamental theoretical and empirical equations that are developed in soil mechanics can be properly used in practice if, and only if, the physical parameters used in those equations are properly evaluated in the laboratory. So, learning to perform laboratory tests of soils plays an important role in the geotechnical engineering profession.

Use of Equipment

Laboratory equipment is never cheap, but the cost can vary widely. For good experimental results, the equipment should be properly maintained. The calibration of certain equipment, such as balances and proving rings, should be checked from time to time. It is always necessary to see that all equipment is clean both before and after use. Better results can be obtained when the equipment being used is clean, so always maintain the equipment as if it were your own.

Recording the Data

In any experiment, it is always a good habit to record all data in the proper table immediately after it has been taken. Oftentimes, students scribble on scratch paper which later may be lost or be illegible. This may lead to conducting the experiment over again or to obtaining inaccurate results.

Report Preparation

For classroom work, most experiments described here may need to be conducted in small groups. However, a report should be written by each student individually. This is a way for students to improve their skills in technical writing. Each report should contain:

1. Cover page—This should include the title of the experiment, name, and date on which the experiment was performed.

2. Following the cover page, the items listed below should be presented in the body of the report:
   (a) Purpose of the experiment
   (b) Equipment used
   (c) A schematic diagram of the main equipment used
   (d) A brief description of the test procedure

3. Results—This should include the data sheet(s), sample calculation(s), and the required graph(s).

4. Conclusion—A discussion of the accuracy of the test procedure; any possible sources of error should be included here.

Comments on Graphs and Tables Prepared for the Report

Graphs and tables should be prepared as neatly as possible. *Always* give the units. Graphs should be made as large as possible, and they should be properly labeled. Examples of a bad graph and a good graph are shown in Fig. 1–1. French curves and a straight edge should be used in preparing graphs when necessary.

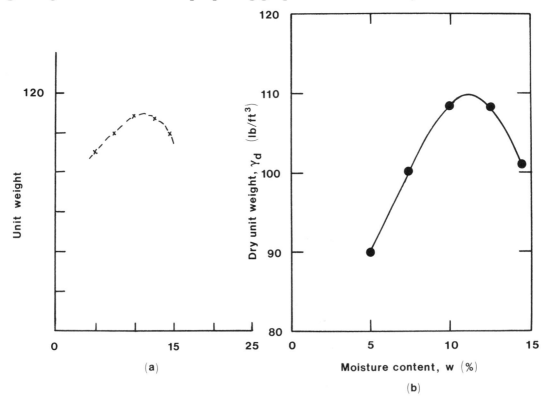

Figure 1-1. (a) A poorly drawn graph for dry unit of soil vs. moisture content. (b) The results given in (a), drawn in a more presentable manner.

Units

The results of laboratory test may be required to be expressed in a given system of units. In the United States, at this time, both English and the SI system of units are used. Conversion of units may be necessary in the preparation of reports. Some selected conversion factors (from English to the SI system) are given in Table 1–1.

Table 1–1. Conversion Factors

	Conventional English Unit	SI Units
Length	1 in.	25.4 mm
	1 ft	0.3048 m
		304.8 mm
Area	1 in.^{2}	$6.4516 \times 10^{-4} m^2$
		6.4516 cm^2
		645.16 mm^2
	1 ft^2	$929 \times 10^{-4} m^2$
		929.03 cm^2
		92903 mm^2
Volume	1 in.^{3}	16.387 cm^3
	1 ft^3	0.028317 m^3
	1 ft^3	28.3168 l
Velocity	1 ft/sec	304.8 mm/sec
		0.3048 m/sec
	1 ft/min	5.08 mm/sec
		0.00508 m/sec
Force	1 lb	4.448 N
Stress	1 lb/in.^{2}	6.9 kN/m^2
	1 lb/ft^2	47.88 N/m^2
Unit Weight	1 lb/ft^3	157.06 N/m^3
Coefficient of Consolidation	1 in.^{2}/sec	6.452 cm^2/sec
	1 ft^2/sec	929.03 cm^2/sec

Standard Test Procedure

In the United States, most laboratories conducting tests on soils for engineering purposes follow the procedures outlined by the American Society for Testing and Materials (ASTM) and the American Association of State Highway and Transportation Officials (AASHTO). The procedures and equipment for soil tests may vary slightly from laboratory to laboratory, but the basic concepts remain the same. The test procedures described in this manual may not be exactly the same as specified by ASTM or AASHTO. However, for students, it is beneficial to know the standard test designations and compare them with the laboratory work actually done. For this reason, some selected AASHTO and ASTM standard test designations are given in Table 1–2.

Table 1–2. Some Important AASHTO and ASTM Soil Test Designations

Name of Test	AASHTO Test Designation	ASTM Test Designation
Moisture content	T-265	D-2216
Specific gravity	T-100	D-854
Sieve analysis	T-87, T-88	D-421
Hydrometer analysis	T-87, T-88	D-422
Liquid limit	T-89	D-4318
Plastic limit	T-90	D-4318
Shrinkage limit	T-92	D-427
Standard Proctor compaction	T-99	D-698
Modified Proctor compaction	T-180	D-1557
Field density by sand cone	T-191	D-1556
Permeability of granular soil	T-215	D-2434
Consolidation	T-216	D-2435
Direct shear (granular soil)	T-236	D-3080
Unconfined compression	T-208	D-2166
Triaxial	T-234	D-2850
AASHTO Soil Classification System	M-145	D-3282
Unified Soil Classification System	—	D-2487

2
Determination of Moisture Content

Introduction

Most laboratory tests in soil mechanics require the determination of moisture content. Moisture content is defined as

$$w = \frac{\text{weight of water present in a given soil mass}}{\text{weight of dry soil}} \qquad (2.1)$$

The moisture content is usually expressed in percent.

For better results, the *minimum* size of the moist soil samples should be approximately as given in Table 2–1.

Table 2–1. Size of Moist Soil Samples for Moisture Content Determination

Maximum Particle Size in the Soil	Minimum Size of Soil Sample (g)
2 in. (50.8 mm)	1000–1200
½ in. (12.7 mm)	300–500
4.75 mm[a]	100–150
0.42 mm[b]	10–40

[a]Size of No. 4 U.S. sieve
[b]Size of No. 40 U.S. sieve

Equipment

1. Moisture can(s).

   These are available in various sizes [for example, 2-in. diameter and $\frac{7}{8}$ in. high (50.8 mm diameter and 22.2 mm high); 3.5-in. diameter and 2 in. high (88.9 mm diameter and 50.8 mm high)].

2. Oven with temperature control.

   For drying, the temperature of the oven is generally kept between 105°C to 110°C. A higher temperature should be avoided. This is to avoid the burning of organic material in the soil and also possible change in the composition of the mineral grains.

Procedure

1. Determine the weight of the empty moisture can plus its cap (W_1), and also record the can number.
2. Place a sample of representative moist soil in the can. Close the can with its cap to avoid loss of moisture.
3. Determine the combined weight of the closed can and wet soil (W_2).
4. Remove the cap from the top of the can and place it on the bottom (of the can).
5. Put the can (Step 4) in the oven for drying the soil to a constant weight. About 24 hours of drying is enough in most cases.
6. Determine the combined weight of the dry soil sample plus the can and its cap (W_3).

Calculation

$$\text{Moisture content (\%)} = \frac{W_2 - W_3}{W_3 - W_1} \times 100 \tag{2.2}$$

A sample calculation of moisture content is given in Table 2–2.

General Comments

Most natural soils, which are sandy or gravelly in nature, may have moisture contents up to about 15 to 20%. In natural fine-grained (silty or clayey) soils, moisture contents up to about 50 to 80% can be found. However, peats and highly organic soils with moisture contents up to 500% are not uncommon.

Following are typical values of moisture content for various types of natural soils.

Soil	Natural Moisture Content in a Saturated State (%)
Loose uniform sand	25–30
Dense uniform sand	12–16
Loose angular-grained silty sand	25
Dense angular-grained silty sand	15
Stiff clay	20
Soft clay	30–50
Soft organic clay	80–130
Glacial till	10

DETERMINATION OF MOISTURE CONTENT

Description of soil _Silty clay_

Location _South Violet Lane_

Sample No. _4_

Can No.	Weight of can, W_1 (g)	Weight of can + wet soil, W_2 (g)	Weight of can + dry soil, W_3 (g)	$w(\%) = \dfrac{W_2 - W_3}{W_3 - W_1} \times 100$
46	16.09	31.32	29.28	15.47

Table 2–2

3
Specific Gravity

Introduction

By specific gravity of soil, we generally refer to the specific gravity of mineral grains (soil solids). Specific gravity, G_s, is defined as

$$G_s = \frac{\text{unit weight of soil solids only}}{\text{unit weight of water}} \tag{3.1}$$

The general ranges of the values of G_s for various soils are given in Table 3–1. The procedure for determination of specific gravity, G_s, described here is applicable for soils composed of *particles smaller than 4.75 mm* (No. 4 U.S. sieve) *in size.*

Table 3–1. General Ranges of G_s for Various Soils

Soil Type	Range of G_s
Sand	2.63–2.67
Silts	2.65–2.7
Clay and silty clay	2.67–2.9
Organic soil	less than 2

Equipment

1. Volumetric flask (500 ml)
2. Thermometer
3. Balance sensitive up to 0.1 g
4. Distilled water
5. Bunsen burner and a stand (and/or vacuum pump or aspirator)

6. Evaporating dishes
7. Spatula
8. Plastic squeeze bottle
9. Drying oven

The equipment required for this experiment is shown in Fig. 3–1.

Procedure

1. Clean the volumetric flask well and dry it.
2. Carefully fill the flask with de-aired, distilled water up to the 500 ml mark (bottom of the meniscus should be at the 500 ml mark).
3. Determine the weight of the flask and the water filled to the 500 ml mark (W_1).
4. Insert the thermometer into the flask with the water and determine the temperature of the water $T = T_1°C$.
5. Put *approximately* 100 grams of air dry soil into an evaporating dish.
6. If the soil is cohesive, add water (de-aired and distilled) to the soil and mix it to the form of a smooth paste. Keep it soaked for about one-half to one hour in the evaporating dish. (*Note:* This step is not necessary for granular, i.e., noncohesive, soils.)
7. Transfer the soil (if granular) or the soil paste (if cohesive) into the volumetric flask.
8. Add distilled water to the volumetric flask containing the soil (or the soil paste) to make it about two-thirds full.
9. Remove the air from the soil-water mixture. This can be done by:
    (a) Gently boiling the flask containing the soil-water mixture for about 15 to 20 minutes. Accompany the boiling with continuous agitation of the flask.

Figure 3–1. Equipment for conducting specific gravity test.

If too much heat is applied, the soil may boil over. Or

(b) Apply vacuum by a vacuum pump or aspirator until all the entrapped air is out.

This is an *extremely important step*. Most of the errors in the results of this test are *due to entrapped air which is not removed*.

10. Bring the temperature of the soil-water mixture in the volumetric flask down to room temperature (i.e., $T_1°C$—see Step 4. This temperature of the water is at room temperature).

11. Add de-aired, distilled water to the volumetric flask until the bottom of the meniscus touches the 500 ml mark. Also dry the outside of the flask and the inside of the neck above the meniscus.

12. Determine the combined weight of the bottle plus soil plus water (W_2).

13. Just as a precaution, check the temperature of the soil and water in the flask to see if it is $T_1°C \pm 1°C$ or not.

14. Pour the soil and water into an evaporating dish. Using a plastic squeeze bottle, wash the inside of the flask and make sure that no soil is left inside.

15. Put the evaporating dish in an oven to dry to a constant weight.

16. Determine the weight of the dry soil in the evaporating dish (W_3).

Calculation

$$G_s = \frac{\text{weight of soil}}{\text{weight of equal volume of water}}$$

where weight of soil $= W_3$
weight of equal volume of water $= (W_1 + W_3) - W_2$

So

$$G_{s \, (\text{at } T_1°C)} = \frac{W_3}{(W_1 + W_3) - W_2} \tag{3.2}$$

Specific gravity is generally reported on the value of the unit weight of water at 20°C. So

$$G_{s \, (\text{at } 20°C)} = G_{s \, (\text{at } T_1°C)} \left(\frac{\gamma_{w \, (\text{at } T_1°C)}}{\gamma_{w \, (\text{at } 20°C)}} \right) \tag{3.3}$$

$$= G_{s \, (\text{at } T_1°C)} A \tag{3.4}$$

where $A = \dfrac{\gamma_{w \, (\text{at } T_1°C)}}{\gamma_{w \, (\text{at } 20°C)}}$

γ_w = unit weight of water

The values of A are given in Table 3–2.

A sample calculation for specific gravity is shown in Table 3–3.

Table 3-2. Values of A [Eq. (3.4)]

Temperature, T_1 (°C)	A
18	1.004
19	1.002
20	1.000
22	0.9996
24	0.9991
26	0.9986
28	0.9980

General Comments

At least two to three specific gravity tests should be conducted. For correct results, these values should not vary by more than 2 to 3%.

SPECIFIC GRAVITY

Description of soil _Light brown sandy silt_

Location _Kent Drive_

Sample No. _23_

Volume of flask at 20°C _500 ml_

Test No.	1		
Volumetric flask no.	6		
Weight of flask + water filled to mark, W_1 (g)	660		
Weight of flask + soil + water filled to mark, W_2 (g)	722		
Weight of dry soil, W_3 (g)	99		
Temperature of test, T_1°C	23		
$G_{s\ (at\ T_1°C)} = \dfrac{W_3}{(W_1+W_3)-W_2}$	2.676		
A	0.9993		
$G_{s\ (at\ 20°C)} = G_{s\ (at\ T_1°C)} \times A$	2.674 ≈2.67		

Table 3-3

4

Sieve Analysis

Introduction

In order to classify a soil for engineering purposes, one needs to know the distribution of the size of grains in the soil mass. Sieve analysis is a method to determine the grain-size distribution of soils. Table 4–1 gives a partial list of the U.S. standard sieve numbers with their corresponding size of openings. Note that, as the sieve number increases, the size of the openings decreases. The sieves are made out of woven wires; those commonly used have a diameter of 8 in. (203 mm). A stack of sieves is shown in Fig. 4–1.

The method of sieve analysis described here is applicable for soils that are *mostly granular with some or no fines.*

Figure 4–1. A stack of sieves with a pan at the bottom and a cover on the top.

Equipment

1. Sieves, a bottom pan, and a cover

   Sieve numbers 4, 10, 20, 40, 60, 140, and 200 are generally used for most standard sieve analysis work

2. A balance sensitive up to 0.1 g
3. Mortar and rubber-tipped pestle
4. Oven
5. Mechanical sieve shaker

Table 4–1. U.S. Sieve Sizes

Sieve No.	Opening (mm)
4	4.75
6	3.35
8	2.36
10	2.00
12	1.68
16	1.18
20	0.85
30	0.60
40	0.425
50	0.300
60	0.250
80	0.180
100	0.150
140	0.106
200	0.075
270	0.053

Figure 4–2. Washing of the soil retained on No. 200 sieve.

Procedure

1. Collect a *representative* oven dry soil sample. Samples having largest particles of the size of No. 4 sieve openings (4.75 mm) should be about 500 grams. For soils having largest particles of size greater than 4.75 mm, larger weights are needed.

2. Break the soil sample into individual particles by using a mortar and a rubber-tipped pestle. (*Note:* The idea is to break up the soil into individual particles, not to break the particles themselves.)

3. Weigh the sample accurately to 0.1 g (W).

4. Make a stack of sieves. A sieve with larger openings is placed above a sieve with smaller openings. The sieve at the bottom should be No. 200. A bottom pan should be placed under sieve No. 200. As mentioned before, generally, the sieves that are used in a stack are Nos. 4, 10, 20, 40, 60, 140, and 200; however, more sieves can be placed in between.

5. Pour the soil prepared in Step 2 into the stack of sieves from the top.

6. Put the cover on the top of the stack of sieves.

7. Run the stack of sieves through a sieve shaker for about 10 to 15 minutes.

8. Stop the sieve shaker and remove the stack of sieves.

9. Weigh the amount of soil retained on each sieve and the bottom pan.

10. If a *considerable amount* of soil with silty and clayey fractions is retained on the No. 200 sieve, it has to be washed. Washing is done by taking the No. 200 sieve with the soil retained on it and pouring water through the sieve from a tap in the laboratory (Fig. 4–2).

Figure 4–3. Back washing to transfer the soil retained on No. 200 sieve to an evaporating dish.

When the water passing through the sieve is clean, stop the flow of water. Transfer the soil retained on the sieve at the end of washing to a porcelain evaporating dish by back washing (Fig. 4–3). Put it in the oven to dry to a constant weight. (*Note:* This step is not necessary if the amount of soil retained on the No. 200 sieve is small.)

Determine the weight of the dry soil retained on No. 200 sieve. The difference between this weight and that retained on No. 200 sieve as determined in Step 9 is the amount of soil that has been washed through.

Calculation

1. Percent of soil retained on n*th* sieve (counting from the top)

$$= \frac{\text{weight retained}}{\text{total weight, } W \text{ (Step 3)}} \times 100 = R_n \qquad (4.1)$$

2. Cumulative percent of soil retained on the n*th* sieve

$$= \sum_{i=1}^{i=n} R_n \qquad (4.2)$$

3. Cumulative percent passing through the n*th* sieve

$$= percent\ finer = 100 - \sum_{i=1}^{i=n} R_n \qquad (4.3)$$

Note: If soil retained on No. 200 sieve is washed, the dry weight determined after washing (Step 10) should be used for calculation of percent finer (than No. 200 sieve). The weight lost due to washing should be added to the weight of the soil retained on the pan.

A sample calculation of sieve analysis is shown in Table 4–2.

Graphs

Plot a graph of percent finer vs. sieve opening on *semilog* graph paper. Sieve opening is plotted on the log scale. (See Fig. 4–4, which is a plot for the calculation shown in Table 4–2.)

Other Calculations

1. Determine D_{10}, D_{30}, and D_{60} (from Fig. 4–4), which are, respectively, the diameters corresponding to percents finer of 10%, 30%, and 60%.

2. Calculate the uniformity coefficient (C_u) and the coefficient of gradation (C_c), using the following equations:

$$C_u = \frac{D_{60}}{D_{10}}$$

$$C_c = \frac{D_{30}^2}{D_{60} \times D_{10}}$$

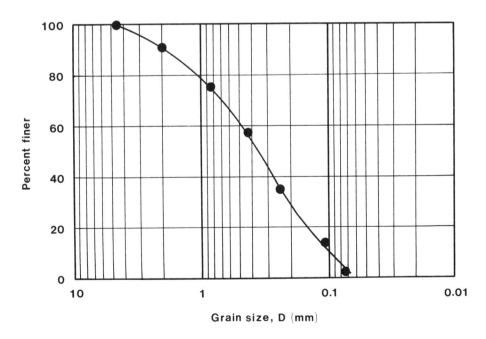

Figure 4–4. Plot of percent finer vs. grain size from the calculation shown in Table 4–2.

As an example, from Fig. 4–4, $D_{60} = 0.46$ mm, $D_{30} = 0.21$ mm, and $D_{10} = 0.098$ mm. So

$$C_u = \frac{0.46}{0.098} = 4.69$$

and

$$C_c = \frac{(0.21)^2}{(0.46)(0.098)} = 0.98$$

General Comments

The diameter D_{10} is generally referred to as *effective size*. The parameters C_u and C_c are generally used for classification of granular soils. The parameter C_c is also referred to as the *coefficient of curvature*. For sand, if C_u is greater than 6 and C_c is between 1 and 3, it is considered well-graded. However, for a gravel to be well-graded, C_u should be greater than 4 and C_c must be between 1 and 3.

The effective size, D_{10}, has been used for several empirical correlations, such as the coefficient of permeability.

SIEVE ANALYSIS

Description of soil *Sand with some fines*

Location *Poplar Street*

Sample No. *2*

Weight of oven dry sample, W *500* (g)

Sieve No.	Sieve opening (mm)	Weight retained on each sieve (g)	Percent of weight retained on each sieve	Cumulative percent retained	Percent finer
4	4.75	0	0	0	100
10	2.0	40.2	8.04	8.04	91.96
20	0.850	84.6	16.92	24.96	75.04
40	0.425	90.2	18.04	43.00	57.00
60	0.250	106.4	21.28	64.28	35.72
140	0.106	108.8	21.76	86.04	13.96
200	0.075	59.4	11.88	97.92	2.08
Pan	——	8.7			

$$\Sigma \quad 498.3 \quad = W_1$$

Loss during sieve analysis $= \dfrac{W - W_1}{W} \times 100 = $ _____ *0.34* _____ % (OK if less than 2%)

Table 4–2

5

Hydrometer Analysis

Introduction

Hydrometer analysis is the procedure generally adopted for determination of the particle-size distribution in a soil for the fraction that is finer than No. 200 sieve size (0.075 mm). The lower limit of the particle-size determined by this procedure is about 0.001 mm.

In hydrometer analysis, a soil specimen is dispersed in water. In a dispersed state in the water, the soil particles will settle individually. It is assumed that the soil particles are spheres, and the velocity of the particles can be given by Stoke's law as

$$v = \frac{\gamma_s - \gamma_w}{18\eta} D^2 \tag{5.1}$$

where v = velocity (cm/sec)

γ_s = specific weight of soil solids (g/cm^3)

γ_w = unit weight of water (g/cm^3)

η = viscosity of water $\left(\dfrac{g \cdot sec}{cm^2} \right)$

D = diameter of the soil particle

If a hydrometer is suspended in water in which soil is dispersed (Fig. 5–1), it will measure the specific gravity of the soil-water suspension at a depth L. The depth L is called the *effective depth*. So, at a time t minutes from the beginning of the test, the soil

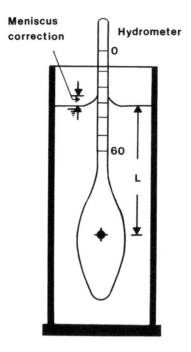

Figure 5-1.
Hydrometer suspended in
water in which the soil is
dispersed.

particles that settle beyond the zone of measurement (i.e., beyond the effective depth, L) will have a diameter given by

$$\frac{L\,(\text{cm})}{t\,(\text{min}) \times 60} = \frac{(\gamma_s - \gamma_w)\,\text{g/cm}^3}{18\eta\,\dfrac{\text{g}\cdot\text{sec}}{\text{cm}^2}}\left[\frac{D\,(\text{mm})}{10}\right]^2$$

$$D\,(\text{mm}) = \frac{10}{\sqrt{60}}\sqrt{\frac{18\eta}{(\gamma_s - \gamma_w)}}\sqrt{\frac{L}{t}} = A\sqrt{\frac{L\,(\text{cm})}{t\,(\text{min})}} \qquad (5.2)$$

$$\text{where} \quad A = \sqrt{\frac{1800\eta}{60\,(\gamma_s - \gamma_w)}} = \sqrt{\frac{30\eta}{(\gamma_s - \gamma_w)}} \qquad (5.3)$$

In the test procedure described here, the *ASTM 152-H* type of hydrometer will be used. From Fig. 5-1 it can be seen that, based on the hydrometer reading (which increases from zero to 60 in the *ASTM 152-H* type of hydrometer), the value of L will change. These calibrated values of L (cm) are shown in Fig. 5-2a and b for various values of hydrometer readings.

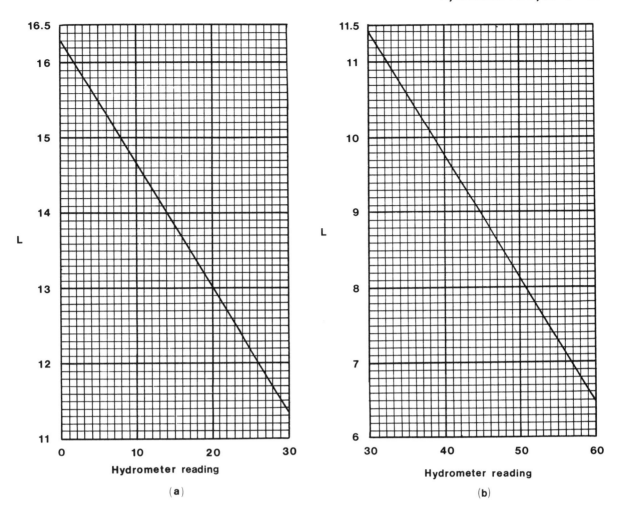

Figure 5-2. Plot of hydrometer reading (*ASTM 152-H* type hydrometer) vs. effective length, *L*.

For actual calculation purposes we also need to know the values of A given by Equation (5.3). An example of this calculation is shown below.

$$\gamma_s = G_s \gamma_w$$

where G_s = specific gravity of soil solids

Thus,

$$A = \sqrt{\frac{30\eta}{(G_s - 1)\gamma_w}} \qquad (5.4)$$

For example, if the temperature of the water is 25°,

$$\eta = 0.0911 \times 10^{-4} \frac{\text{g} \cdot \text{sec}}{\text{cm}^2}$$

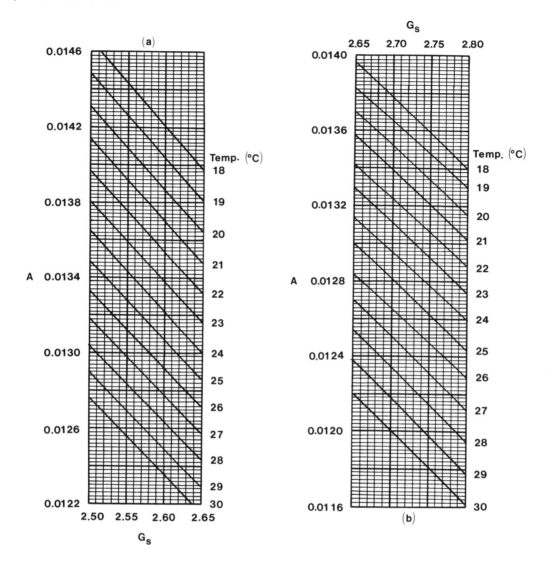

Figure 5–3. Variation of A with G_s and temperature.

and $G_s = 2.7$

$$A = \sqrt{\frac{30\,(0.0911 \times 10^{-4})}{(2.7 - 1)\,(1)}} = 0.01267$$

The variations of A with G_s and the water temperature are shown in Fig. 5–3a and b.

The *ASTM 152-H* type of hydrometer is calibrated up to a reading of 60 at a temperature of 20°C for soil particles having a $G_s = 2.65$. A hydrometer reading of, say, 30 at a given time of a test means that there are 30 g of soil solids ($G_s = 2.65$) in suspension per 1000 cc of soil-water mixture at a temperature of 20°C at a depth where the specific gravity of the soil-water suspension is measured (i.e., L). From this measurement, we can determine the percentage of soil still in suspension at time t from the beginning of the test, and all the soil particles will have diameters smaller than D calculated by Equation (5.2). However, in the actual experimental work, some corrections to the observed hydrometer readings need to be applied. They are as follows:

1. Temperature correction (F_T) — The actual temperature of the test may not be 20°C. The temperature correction (F_T) can be approximated as

$$F_T = -4.85 + 0.25T \quad \text{(for } T \text{ between } 15° \text{ and } 28°\text{C)} \tag{5.5}$$

   where F_T = temperature correction to the observed reading (can be either positive or negative)

   T = temperature of test in °C

2. Meniscus correction (F_m) — Generally, the upper level of the meniscus is taken as the reading during laboratory work (F_m is always positive).

3. Zero correction (F_z) — A deflocculating agent is added to the soil-distilled water suspension for performing experiments. This will change the zero reading (F_z can be positive or negative).

Equipment

1. ASTM 152-H hydrometer
2. Mixer
3. Two 1000 cc graduated cylinders
4. Thermometer
5. Constant temperature bath
6. Deflocculating agent
7. Spatula
8. Beaker
9. Balance
10. Plastic squeeze bottle
11. Distilled water
12. No. 12 rubber stopper

The equipment necessary (except the balance and the constant temperature bath) is shown in Fig. 5–4.

Procedure

Note: This procedure is used when more than 90 percent of the soil is finer than No. 200 sieve.

1. Take 50 g of oven-dry, well-pulverized soil in a beaker.
2. Prepare a deflocculating agent. Usually, a 4% solution of sodium hexametaphosphate (Calgon) is used. This can be prepared by adding 40 g of Calgon in 1000 cc of distilled water and mixing it thoroughly.

Figure 5–4. Equipment for hydrometer test.

3. Take 125 cc of the mixture prepared in Step 2, and add it to the soil taken in Step 1. This should be allowed to soak for about 8 to 12 hours.

4. Take a 1000 cc graduated cylinder, and add 875 cc of distilled water plus 125 cc of deflocculating agent in it. Mix the solution well.

5. Put the cylinder (from Step 4) in a constant temperature bath. Record the temperature of the bath, T (in °C).

6. Put the hydrometer in the cylinder (Step 5). Record the reading. (*Note:* The *top of the meniscus* should be read.) This is the zero correction (F_z), which can be +ve or −ve. Also observe the meniscus correction (F_m).

7. Using a spatula, mix thoroughly the soil prepared in Step 3. Pour it into the mixer cup.

   Note: During this process, some soil may stick to the side of the beaker. Using the plastic squeeze bottle filled with distilled water, wash all the remaining soil in the beaker into the mixer cup.

8. Add distilled water to the cup to make it about two-thirds full. Mix it for about two minutes using the mixer.

9. Pour the mix into the second graduated 1000 cc cylinder. Make sure that all of the soil solids are washed out of the mixer cup. Fill the graduated cylinder with distilled water to bring the water level up to the 1000 cc mark.

10. Secure a No. 12 rubber stopper on the top of the cylinder (Step 9). Mix the soil-water well by turning the soil cylinder upside down several times.

11. Put the cylinder into the constant temperature bath next to the cylinder described in Step 5. Record the time immediately. This is cumulative time $t = 0$. Insert the hydrometer into the cylinder containing the soil-water suspension.

12. Take hydrometer readings at cumulative times $t = 0.25$ min., 0.5 min., 1 min., and 2 min. Always read the upper level of the meniscus.

13. Take the hydrometer out after two minutes, and put it into the cylinder next to it (Step 5).

14. Hydrometer readings are to be taken at time $t = 4$ min., 8 min., 15 min., 30 min., 1 hr., 2 hr., 4 hr., 8 hr., 24 hr., and 48 hr. For each reading, insert the hydrometer into the cylinder containing the soil-water suspension about 30 seconds before the reading is due. After the reading is taken, remove the hydrometer and put it back into the cylinder next to it (Step 5).

Calculation

Refer to Table 5–1.

Column 2 — These are observed hydrometer readings (R) corresponding to times given in Column 1.

Column 3 — R_{cp} = corrected hydrometer reading for calculation of percent finer

$$= R + F_T - F_z \tag{5.6}$$

Column 4 — *Percent finer* $= \dfrac{a \, R_{cp}}{W_s} (100)$

where W_s = dry weight of soil used for the hydrometer analysis
a = correction for specific gravity (since the hydrometer is calibrated for $G_s = 2.65$)

$$= \dfrac{G_s (1.65)}{(G_s - 1) \, 2.65} \qquad \text{(See Fig. 5–5)} \tag{5.7}$$

Column 5 — R_{cL} = corrected reading for determination of effective length

$$= R + F_m \tag{5.8}$$

Column 6 — Determine L (effective length) corresponding to the values of R_{cL} (Col. 5) given in Fig. 5–2.

Column 7 — Determine A from Fig. 5–3.

Column 8 — Determine

$$D \, (\text{mm}) = A \sqrt{\dfrac{L \, (\text{cm})}{t \, (\text{min})}}$$

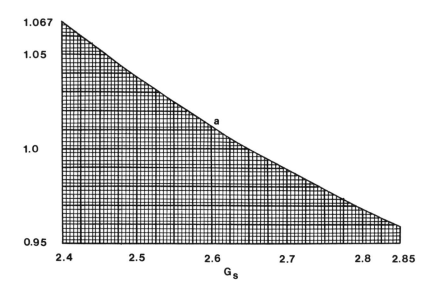

Figure 5-5. Variation of a with G_s (Eq. 5.7).

Graph

Plot a grain-size distribution graph on semilog graph paper with percent finer (Col. 4, Table 5–1) on the natural scale and D (Col. 8, Table 5–1) on the log scale.

A sample calculation and the corresponding graph are shown in Table 5–1 and Fig. 5–6, respectively.

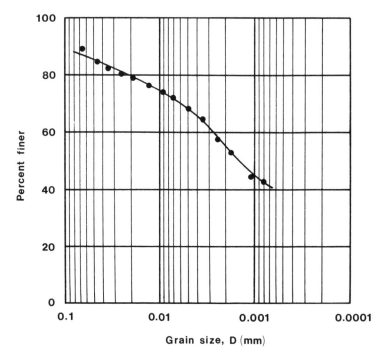

Figure 5-6. Plot of percent finer vs. grain size from the results given in Table 5–1.

Hydrometer Analysis / 31

HYDROMETER ANALYSIS

Description of soil _Brown silty clay_

Sample No._____ Location_____

G_s _____2.75_____ Hydrometer type __ASTM 152-H_____

Dry weight of soil, W_s_____50g____ (g) Temperature of test, T__28_____ (°C)

Meniscus correction, F_m _____1_____ Zero correction, F_z_____+7_____

Temperature correction, F_T____+2.15_____

Time (min)	Hydrometer reading, R	R_{cp}	Percent finer, $\frac{aR_{cp}}{50} \times 100$	R_{cL}	L (cm)	A	D (mm)
(1)	(2)	(3)	(4)	(5)	(6)	(7)	(8)
0.25	51	46.15	90.3	52	7.8	0.0121	0.068
0.5	48	43.15	84.4	49	8.3		0.049
1	47	42.15	82.4	48	8.4		0.035
2	46	41.15	80.5	47	8.6		0.025
4	45	40.15	78.5	46	8.8		0.018
8	44	39.15	76.6	45	8.95		0.013
15	43	38.15	74.6	44	9.1		0.009
30	42	37.15	72.7	43	9.25		0.007
60	40	35.15	68.8	41	9.6		0.005
120	38	33.15	64.8	39	9.9		0.0035
240	34	29.15	57.0	35	10.5		0.0025
480	32	27.15	53.1	33	10.9		0.0018
1440	29	24.15	47.23	30	11.35		0.0011
2880	27	22.15	43.3	28	11.65		0.0008

Table 5–1

Procedure Modification

When a smaller amount (less than about 90%) of soil is finer than No. 200 sieve size, the following modification to the above procedure needs to be applied.

1. Take an oven-dry specimen of soil. Determine its weight (W_1).
2. Pulverize the soil using a mortar and rubber-tipped pestle, as described in Chapter 4.
3. Run a sieve analysis on the soil (Step 2), as described in Chapter 4.
4. Collect in the bottom pan the soil passing through No. 200 sieve.
5. Wash the soil retained on No. 200 sieve, as described in Chapter 4. Collect all the wash water, and dry it in an oven.
6. Mix together the minus No. 200 portion from Step 4 and the dried minus No. 200 portion from Step 5.
7. Calculate the percent finer for the soil retained on No. 200 sieve and above (as shown in Table 4–1).
8. Take 50 g of the minus No. 200 soil (Step 6) and run a hydrometer analysis. (Follow Steps 1 through 14 as described previously.)
9. Report the calculations for the hydrometer analysis similar to that shown in Table 5–1. However, note that the percent finer now calculated (as in Col. 8 of Table 5–1) is *not the percent finer based on the total sample*. Calculate the percent finer based on the total sample as

$$P_T = (\text{Col. 8 of Table 5–1}) \left(\frac{\text{percent passing No. 200 sieve}}{100} \right)$$

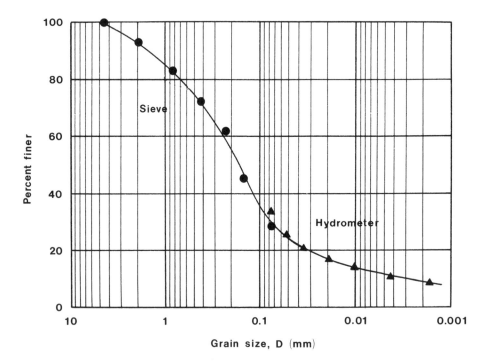

Figure 5–7. A grain-size distribution plot—combined results from sieve analysis and hydrometer analysis.

Percent passing No. 200 sieve can be obtained from Step 7 above.

10. Plot a combined graph for percent finer vs. grain-size distribution obtained from *both the sieve analysis and the hydrometer analysis*. An example of this is shown in Fig. 5–7. From this plot, note that there is an overlapping zone. The percent finer calculated from the sieve analysis for a given grain size does not match that calculated from the hydrometer analysis. This is due to the different assumptions used in these two experiments. The grain sizes obtained from a sieve analysis are the least sizes of soil grains, and the grain sizes obtained from the hydrometer analysis are the diameters of equivalent spheres of soil grains.

General Comments

A hydrometer analysis gives results from which the percent of soil finer than 0.002 mm in diameter can be estimated. It is generally accepted that the percent finer than 0.002 mm in size is clay or clay-size fractions. Most clay particles are smaller than 0.001 mm, and 0.002 mm is the upper limit. The presence of clay in a soil contributes to its plasticity.

6
Liquid Limit Test

Introduction

When a *cohesive soil* is mixed with an excessive amount of water, it will be in a somewhat *liquid state* and flow like a viscous liquid. However, when this viscous liquid is gradually dried, with the loss of moisture, it will pass into a *plastic state*. With further reduction of moisture, the soil will pass into a semisolid and then into a solid state. This is shown in Fig. 6–1. The moisture content (in percent) at which the cohesive soil will pass from a liquid state to a plastic state is called the *liquid limit* of the soil. Similarly, the moisture contents (in percent) at which the soil changes from a plastic to a semisolid state and from a semisolid state to a solid state are referred to as the *plastic limit* and the *shrinkage limit*, respectively. These limits are referred to as the *Atterberg limits* (1911). In this chapter, the procedure for determination of the liquid limit of a cohesive soil will be discussed.

Equipment

1. Casagrande liquid limit device
2. Grooving tool
3. Moisture cans

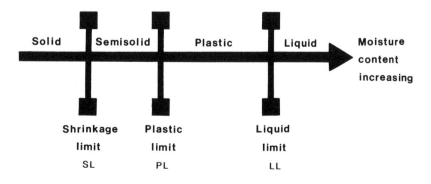

Figure 6–1. Atterberg limits.

4. Porcelain evaporating dish
5. Spatula
6. Oven
7. Balance sensitive up to 0.01 g
8. Plastic squeeze bottle
9. Paper towels

The equipment (except the balance and the oven) is shown in Fig. 6–2.

The Casagrande liquid limit device essentially consists of a brass cup that can be raised and dropped through a distance of 10 mm on a hard rubber base by a cam operated by a crank (see Fig. 6–3a). Fig. 6–3b shows a schematic diagram of a grooving tool.

Procedure

1. Determine the weight of three moisture cans (W_1).
2. Put about 250 g of air-dry soil, passed through No. 40 sieve, into an evaporating dish. Add water from the plastic squeeze bottle, and mix the soil to the form of a uniform paste.
3. Place a portion of the paste in the brass cup of the liquid limit device. Using the spatula, smooth the surface of the soil in the cup such that the maximum depth of the soil is about 8 mm.
4. Using the grooving tool, cut a groove along the center line of the soil pat in the cup (Fig. 6–4a).
5. Turn the crank of the liquid limit device at the rate of about 2 revolutions per second. By this, the liquid limit cup will rise and drop through a vertical distance of 10 mm once for each revolution. The soil from two sides of the cup will

Figure 6–2. Equipment for liquid limit test.

begin to flow toward the center. Count the number of blows, N, for the groove in the soil to close through a distance of $\frac{1}{2}$ in. (12.7 mm), as shown in Fig. 6–4b.

If N = about 25 to 35, collect a moisture sample from the soil in the cup in a moisture can. Close the cover of the can, and determine the weight of the can plus the moist soil (W_2).

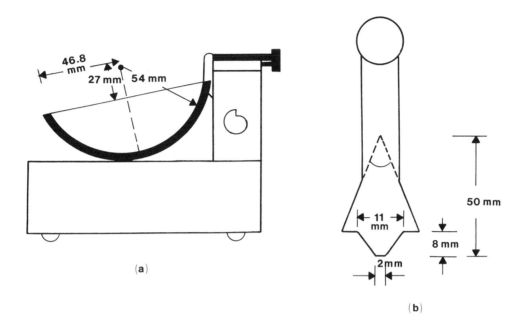

Figure 6–3. Schematic diagram of: (a) liquid limit device, (b) grooving tool.

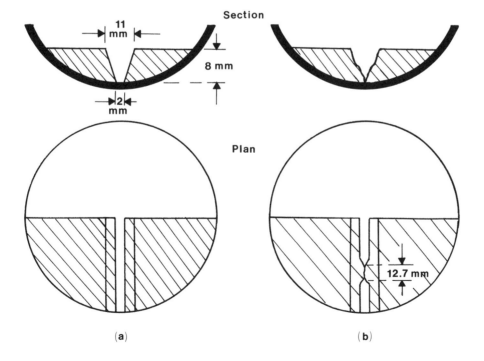

Figure 6–4. Schematic diagram of soil pat in the cup of the liquid limit device at: (a) beginning of test, (b) end of test.

Remove the rest of the soil paste from the cup to the evaporating dish. Use paper towels to thoroughly clean the cup.

If the soil is too dry, N will be more than about 35. In that case, remove the soil by the spatula to the evaporating dish. Clean the liquid limit cup thoroughly with paper towels. Mix the soil in the evaporating dish with more water, and try again.

If the soil is too wet, N will be less than about 25. In that case, remove the soil in the cup to the evaporating dish. Clean the liquid limit cup carefully with paper towels. Stir the soil paste with the spatula for some time to dry it up. The evaporating dish may be placed in the oven for a few minutes for drying also. *Do not add dry soil to the wet-soil paste to reduce the moisture content for bringing it to the proper consistency.* Now try again in the liquid limit device to get the groove closure of $\frac{1}{2}$ in. (12.7 mm) between 25 and 35 blows.

6. Add more water to the soil-paste in the evaporating dish, and mix thoroughly. Repeat Steps 3, 4, and 5 to get a groove closure of $\frac{1}{2}$ in. (12.7 mm) in the liquid limit device at a blow count, $N = 20$ to 25. Take a moisture sample from the cup. Remove the rest of the soil paste to the evaporating dish. Clean the cup with paper towels.

7. Add more water to the soil paste in the evaporating dish, and mix well. Repeat Steps 3, 4, and 5 to get a blow count, N, between 15 and 20 for a groove closure of $\frac{1}{2}$ in. (12.7 mm) in the liquid limit device. Take a moisture sample from the cup.

8. Put the three moisture cans in the oven to dry to constant weights (W_3). (The caps of the moisture cans should be removed from the top and placed at the bottom of the respective cans in the oven.)

Calculation

Determine the moisture content for each of the three trials (Steps 5, 6 and 7) as

$$w\,(\%) \;=\; \frac{W_2 - W_3}{W_3 - W_1}\,(100) \tag{6.1}$$

Graph

Plot a semilog graph between moisture content (arithmetic scale) vs. number of blows, N (log scale). This will approximate a straight line, which is called the *flow curve*. From the straight line, determine the moisture content $w(\%)$ corresponding to 25 blows. This is the *liquid limit* of the soil.

The slope of the flow line is called the *flow index*, F_I, or

$$F_I = \frac{w_1\,(\%) - w_2\,(\%)}{\log\,N_2 - \log\,N_1} \tag{6.2}$$

Typical examples of liquid limit calculation and the corresponding graph are shown in Table 6–1 and Fig. 6–5.

LIQUID LIMIT TEST

Description of soil *Gray silty clay*

Location *Locust Avenue*

Sample No. *4*

Can No.	Weight of can, W_1 (g)	Weight of can + wet soil, W_2 (g)	Weight of can + dry soil, W_3 (g)	Moisture content, w (%)	Number of blows, N
8	15.26	29.30	25.84	32.7	35
21	17.01	31.58	27.72	36.04	23
25	15.17	31.45	26.96	38.1	17

Liquid limit = *35.2*

Flow index = $\dfrac{37 - 33.7}{\log 30 - \log 20} = 18.74$

Table 6–1

General Comments

Based on the liquid limit tests on several soils, the U.S. Army Waterways Experiment Station (1949) has observed that the liquid limit, LL, of a soil can be approximately given by

$$LL = w_N(\%)\left(\frac{N}{25}\right)^{0.121} \tag{6.3}$$

where $w_N(\%)$ = moisture content, in percent, for $\frac{1}{2}$ in. (12.7 mm) groove closure in the liquid limit device at N number of blows

ASTM also recommends this equation for determination of liquid limit of soils (ASTM designation D-4318). However, the value of w_N should correspond to an N value between 20 and 30. Following are the values of $(N/25)^{0.121}$ for various values of N.

N	$\left(\dfrac{N}{25}\right)^{0.121}$
20	0.973
21	0.979
22	0.985
23	0.990
24	0.995
25	1.000
26	1.005
27	1.009
28	1.014
29	1.018
30	1.022

The presence of clay contributes to the plasticity of soil. The liquid limit of a soil will change depending on the amount and type of clay minerals present in it. Following are the approximate ranges for the liquid limit of some clay minerals.

Clay mineral	LL
Kaolinite	35–100
Illite	55–120
Montmorillonite	100–800

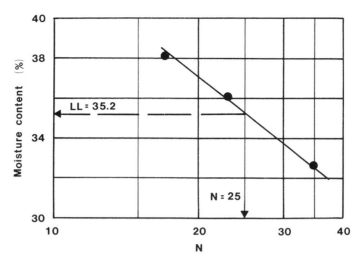

Figure 6-5. Plot of moisture content (%) vs. number of blows for the liquid limit test results reported in Table 6-1.

7

Plastic Limit Test

Introduction

The fundamental concept of *plastic limit* was introduced in the introductory section of the preceding chapter (see Fig. 6–1). It is defined as the moisture content, in percent, at which a cohesive soil will change from a *plastic state* to a *semisolid state*. In the laboratory, the *plastic limit* is defined as the moisture content (%) at which a thread of soil will just crumble when rolled to a diameter of $\frac{1}{8}$-in. (3.18 mm). This test might be seen as somewhat arbitrary and, to some extent, the result might depend on the person performing the test. However, with practice, fairly consistent results may be obtained.

Equipment

1. Porcelain evaporating dish
2. Spatula
3. Plastic squeeze bottle with water
4. Moisture can
5. Ground glass plate
6. Balance sensitive up to 0.01 g

Procedure

1. Put approximately 20 grams of a representative, air-dry soil sample, passed through No. 40 sieve, into a porcelain evaporating dish.
2. Add water from the plastic squeeze bottle to the soil, and mix thoroughly.
3. Take the weight of a moisture can, and record it on the data sheet (W_1).
4. From the moist soil prepared in Step 2, prepare several ellipsoidal-shaped soil masses by squeezing the soil with your fingers.
5. Take one of the ellipsoidal-shaped soil masses (Step 4), and roll it on a ground glass plate using the palm of your hand (Fig. 7–1). The rolling should be done at

43

Figure 7-1. An ellipsoidal soil mass is being rolled to a thread on a glass plate.

the rate of about 80 strokes per minute. Note that one complete backward and one complete forward motion of the palm constitute a stroke.

6. When the thread being rolled in Step 5 reaches $\frac{1}{8}$-in. (3.18 mm) in diameter, break it up into several small pieces and squeeze it with your fingers to form an ellipsoidal mass again.

7. Repeat Steps 5 and 6 until the thread crumbles into several pieces when it reaches a diameter of $\frac{1}{8}$-in. (3.18 mm). It is possible that a thread may crumble at a diameter larger than $\frac{1}{8}$-in. (3.18 mm) during a given rolling process, whereas it did not crumble at the same diameter during the immediately previous rolling.

8. Collect the small crumbled pieces in the moisture can, and close the cover.

9. Take the other ellipsoidal soil masses formed in Step 4, and repeat Steps 5 through 8.

10. Take the combined weight of the moisture can plus the wet soil (W_2). Remove the cap from the top of the can, and place the can (with the cap at the bottom of the can) in the oven.

11. After about 24 hours, remove the can from the oven and take the weight of the can plus the dry soil (W_3).

Calculations

$$\text{Plastic limit} = \frac{\text{weight of moisture}}{\text{weight of dry soil}}(100) \qquad (7.1)$$

$$= \frac{W_2 - W_3}{W_3 - W_1}(100)$$

The results may be presented in a tabular form as shown in Table 7–1. If the liquid limit of the soil is known, calculate the *plasticity index, PI*, as

$$PI = LL - PL \qquad (7.2)$$

General Comments

The liquid limit and the plasticity index of cohesive soils are important parameters for classification purposes. The engineering soil classification systems are described in Appendix A. The plasticity index has also been used to determine the activity of a clayey soil. Activity, *A*, is defined as

$$A = \frac{PI}{(\% \text{ of clay-size fraction, by weight})}$$

Following are typical values of *PI* of several clay minerals:

Clay minerals	PI
Kaolinite	20–40
Illite	35–55
Montmorillonite	50–100

PLASTIC LIMIT TEST

Description of soil *Gray clayey silt*

Location *Locust Avenue*

Sample No. *3*

Can No.	Weight of can, W_1 (g)	Weight of can + wet soil, W_2 (g)	Weight of can + dry soil, W_3 (g)	$PL = \dfrac{W_2 - W_3}{W_3 - W_1} \times 100$
103	13.33	23.86	22.27	17.78

Plasticity index = $PI = LL - PL$ = *34 − 17.78* = *16.28*

Table 7-1

8
Shrinkage Limit Test

Introduction

The fundamental concept of *shrinkage limit* was presented in Fig. 6–1. A saturated clayey soil, when gradually dried, will lose moisture, and, hence, there will be a reduction in the volume of the soil mass. During the drying process, a condition will be reached when any further drying will result in a reduction of moisture content without any decrease of volume (Fig. 8–1). The moisture content of the soil, in percent, at which the decrease in soil volume ceases is defined as the *shrinkage limit*.

Equipment

1. Shrinkage limit dish [usually made of porcelain, about 1.75-in. (44.4 mm) in diameter and 0.5-in. (12.7 mm) high]
2. A glass cup [about 2.25- to 2.5-in. (57.15 to 63.5 mm) in diameter and about 1.25- to 1.5-in. (31.75 to 38.1 mm) high]
3. Glass plate with three prongs
4. Porcelain evaporating dish
5. Spatula
6. Plastic squeeze bottle with water
7. Steel straight edge
8. Mercury
9. Watch glass
10. Balance sensitive to 0.01 g

The above required equipment is shown in Fig. 8–2.

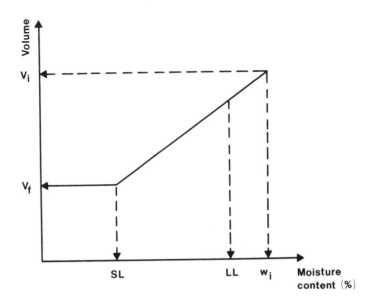

Figure 8-1. Definition of shrinkage limit.

Procedure

1. Put about 80 to 100 grams of a representative air-dry soil, passed through No. 40 sieve, into an evaporating dish.

2. Add water to the soil from the plastic squeeze bottle, and mix it thoroughly into the form of a creamy paste. Note that the moisture content of the paste should be above the liquid limit of the soil to ensure full saturation.

3. Coat the shrinkage limit dish lightly with petroleum jelly, and then determine the weight of the coated dish (W_1).

4. Fill the dish about one-third full with the soil paste. Tap the dish on a firm surface so that the soil flows to the edges of the dish and no air bubbles exist.

5. Repeat Step 4 until the dish is full.

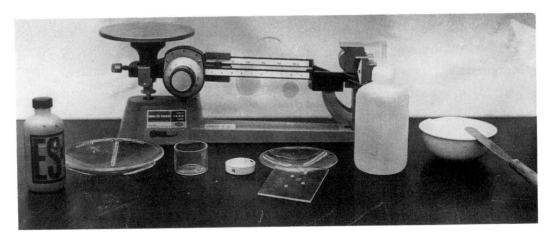

Figure 8-2. Equipment necessary for determination of shrinkage limit.

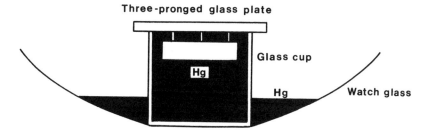

Figure 8-3. Determination of the volume of the soil pat (Step 12).

6. Strike the dish off with the steel straight edge. Clean the sides and the bottom of the dish with paper towels.

7. Determine the weight of the dish plus the wet soil (W_2).

8. Allow the dish to air dry (about 6 hours) until the color of the soil pat becomes lighter. Then put the dish with the soil into the oven to dry.

9. Determine the weight of the dish and the oven-dry soil pat (W_3).

10. Remove the soil pat from the dish.

11. In order to find the volume of the shrinkage limit dish, fill the dish with mercury. (*Note:* The dish should be placed on a watch glass.) Use the three-pronged glass plate, and level the surface of the mercury in the dish. The excess mercury will flow into the watch glass. Determine the weight of mercury in the dish (W_4) in grams.

12. In order to determine the volume of the soil pat, fill the glass cup with mercury. (The cup should be placed on a watch glass.) Using the three-pronged glass plate, level the surface of the mercury in the glass cup. Remove the excess mercury on the watch glass. Place the dry soil pat on the mercury in the glass cup. The soil pat will float. Now, using the three-pronged glass plate, push the soil pat into the mercury slowly until the soil pat is completely submerged (Fig. 8–3). The displaced mercury will flow out of the glass cup and will be collected on the watch glass. Determine the weight of the displaced mercury on the watch glass (W_5) in grams.

Calculation

The initial moisture content of the soil at molding

$$= w_i\,(\%) \;=\; \frac{W_2 - W_3}{W_3 - W_1}\,(100) \tag{8.1}$$

The change in moisture content (%) before the volume reduction ceased (refer to Fig. 8–1)

$$= \frac{(V_i - V_f)\,\gamma_w}{\text{Dry weight of soil pat}} \;=\; \frac{(W_4 - W_5)}{13.6\,(W_3 - W_1)}\,(100) \tag{8.2}$$

Hence,

$$\text{Shrinkage limit} = SL = w_i(\%) - \frac{(W_4 - W_5)}{13.6\,(W_3 - W_1)}\,(100) \tag{8.3}$$

Note that W_4 and W_5 are in grams, and the specific gravity of the mercury is 13.6. A sample calculation is shown in Table 8–1.

General Comments

The ratio of the liquid limit to the shrinkage limit (LL/SL) of a soil gives a good idea about the shrinkage properties of the soil. If the ratio of LL/SL is large, the soil in the field may undergo undesirable volume change due to change of moisture. New foundations constructed on these soils may show cracks due to shrinking and swelling of soil that result from seasonal moisture change.

SHRINKAGE LIMIT TEST

Description of soil *Dark brown clayey*

Location *San Blas Drive*

Sample No. *8*

Test No.	1		
Weight of coated shrinkage limit dish, W_1 (g)	12.34		
Weight of dish + wet soil, W_2 (g)	40.43		
Weight of dish + dry soil, W_3 (g)	33.68		
$w_i = \dfrac{(W_2 - W_3)}{(W_3 - W_1)} \times 100, (\%)$	31.63		
Weight of mercury to fill the dish, W_4 (g)	198.83		
Weight of mercury displaced by soil pat, W_5 (g)	150.30		
$\dfrac{(W_4 - W_5)}{(13.6)\,(W_3 - W_1)}(100)$	16.72		
$SL = w_i - \dfrac{(W_4 - W_5)}{(13.6)\,(W_3 - W_1)}(100)$	14.91		

Table 8-1

9
Constant Head Permeability Test in Sand

Introduction

The rate of the flow of water through a soil specimen of gross cross-sectional area, A, can be expressed as

$$q = kiA \qquad (9.1)$$

where q = flow in unit time
  k = coefficient of permeability
  i = hydraulic gradient.

For coarse sands, the value of the coefficient of permeability may vary from 1 to 0.01 cm/sec., and for fine sand it may be in the range of 0.01 to 0.001 cm/sec.

Several relations between k and the void ratio (e) for sandy soils have been proposed. They are of the form

$$k \propto e^2 \qquad (9.2)$$

$$k \propto \frac{e^2}{1+e} \qquad (9.3)$$

$$k \propto \frac{e^3}{1+e} \qquad (9.4)$$

The coefficient of permeability of sands can be easily determined in the laboratory by two simple methods. They are (a) the constant head test and (b) the variable head test. In this chapter, the *constant head test method* will be described.

Equipment

1. Constant head permeameter
2. Graduated cylinder (250 cc or 500 cc)
3. Balance, sensitive up to 0.1 g
4. Thermometer, sensitive up to 0.1°C
5. Rubber tubing
6. Stop watch

Constant Head Permeameter

A schematic diagram of a constant head permeameter is shown in Fig. 9–1. This can be assembled in the laboratory at very little cost. It essentially consists of a plastic soil specimen cylinder, two porous stones, two rubber stoppers, one spring, one constant head chamber, a large funnel, a stand, a scale, three clamps, and some plastic tubes. The plastic cylinder may have an inside diameter of 2.5 in. (63.5 mm). This is because 2.5 in. (63.5 mm) diameter porous stones are usually available in most soils laboratories. The length of the sample tube may be about 12 in. (304.8 mm).

Procedure

1. Take the weight of the plastic specimen tube, the porous stones, the spring, and the two rubber stoppers (W_1).
2. Slip the bottom porous stone into the specimen tube, and then fix the bottom rubber stopper to the specimen tube.
3. Collect oven-dry sand in a container. Using a spoon, pour the sand into the specimen tube in small layers, and compact it by vibration and/or other compacting means.

   Note: By changing the degree of compaction, we can prepare a number of test specimens having different void ratios.
4. When the length of the specimen is about two-thirds the length of the tube, slip the top porous stone into the tube to rest firmly on the specimen.
5. Place a spring on the top porous stone, if necessary.
6. Fix a rubber stopper to the top of the sample tube.

   Note: The spring in the assembled position will not allow any expansion of the specimen volume, and thus of the void ratio, during the test. The spring is not shown in Fig. 9–1.
7. Determine the weight of the assembly (Step 6 — W_2).
8. Measure the length (L) of the compacted specimen in the tube.
9. Assemble the permeameter near a sink, as shown in Fig. 9–1.
10. Run water into the top of the large funnel fixed to the stand through a plastic tube from the water inlet. The water will flow through the specimen to the constant head chamber. After some time, the water will flow out to the sink through the outlet in the constant head chamber.

    Note: Make sure that water does not leak from the sample tube.

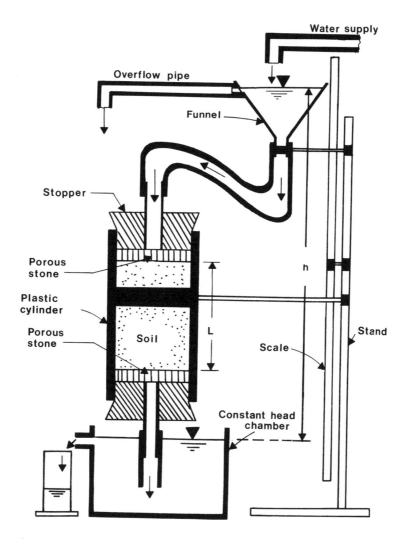

Figure 9-1. Schematic diagram of constant head permeability test setup.

11. Adjust the supply of water to the funnel so that the water level in the funnel remains constant. At the same time, allow the flow to continue for about 10 minutes in order to saturate the specimen.

    Note: Some air bubbles may appear in the plastic tube connecting the funnel to the specimen tube. Remove them.

12. After a steady flow is established (that is, once the head difference h is constant), in a graduated cylinder collect the water flowing out of the constant head chamber (Q). Record the collection time (t) with a stop watch.

13. Repeat Step 12 about three times. Keep the collection time, t, the same, and determine Q. Find the average value of Q.

14. Change the head difference, h, and repeat Steps 11, 12, and 13 about three times.

15. Record the temperature, T, of the water.

Calculation

1. Calculate the void ratio of the compacted specimen as follows:

   Dry unit weight of soil specimen =

   $$\gamma_d = \frac{W_2 - W_1}{\frac{\pi}{4}D^2 L}$$

   Thus

   $$e = \frac{G_s \gamma_w}{\gamma_d} - 1 \qquad (9.5)$$

   where G_s = specific gravity of soil solids
   γ_w = unit weight of water
   D = diameter of the specimen.

2. Calculate k as

   $$k = \frac{QL}{Aht} \qquad (9.6)$$

   where $A = \frac{\pi}{4}D^2$

3. The value k is usually given for a test temperature of water at 20°C. So calculate $k_{20°C}$ as

   $$k_{20°C} = k_{T°C} \frac{\eta_T}{\eta_{20}} \qquad (9.7)$$

   where $\eta_{T°C}$ and $\eta_{20°C}$ are viscosities of water at $T°C$ and 20°C, respectively.

   Fig. 9–2 gives a plot of $\dfrac{\eta_{T°C}}{\eta_{20°C}}$ against T (in °C).

   Table 9–1 gives a sample calculation for the permeability test.

General Comments

Equation 9.1 is based on Darcy's law, which may be expressed as

$$v = ki$$

where v = discharge velocity.

Darcy's law is valid when the flow of water through the pore spaces of the soil is laminar. However, for very coarse sands, gravels, and boulders, a turbulent flow of water can be expected. In such cases Darcy's law is not valid. The hydraulic gradient, i, can be expressed as

$$i = av + bv^2$$

where a and b = constants.

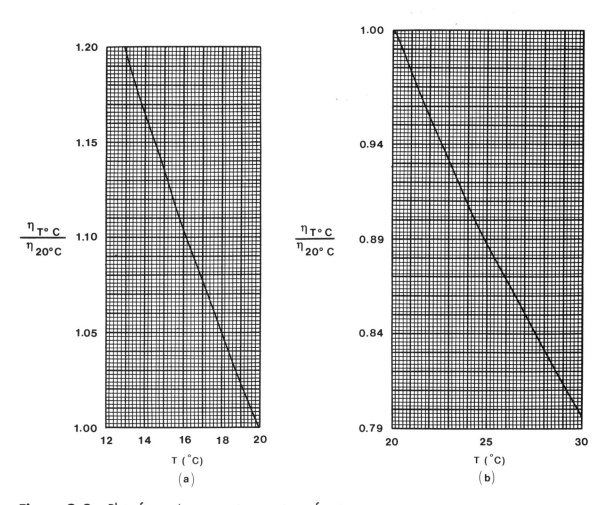

Figure 9–2. Plot of $\eta_{T°C}/\eta_{20°C}$ vs. temperature of water.

CONSTANT HEAD PERMEABILITY TEST

Description of soil _Uniform sand_

Sample No._____ Location_____

Length of specimen, L _13.2_ (cm) Diameter of specimen, D _6.35_ (cm)

Area of specimen, A _31.65_ (cm²) Dry weight of specimen, $W_2 - W_1$ _726.9_ g

Void ratio, e _0.53_ G_s _2.66_

Test No.	Average flow, Q (cm³)	Time of collection, t (sec)	Temperature of water, T (°C)	Head difference, h (cm)	$k = \dfrac{QL}{Aht}$ (cm/sec)
1	305	60	25	60	0.035
2	375	60	25	70	0.037
3	395	60	25	80	0.034

Average k = _0.035_ cm/sec

$$k_{(20°C)} = k_{T°C} \frac{\eta_{T°C}}{\eta_{20°C}} = 0.035(0.89) = 0.0312 \text{ cm/sec}$$

Table 9–1

10
Falling Head Permeability Test in Sand

Introduction

We have discussed the procedure for conducting the constant head permeability tests in sands in the preceding chapter. The falling head permeability test is another experimental procedure for determining the coefficient of permeability of sand.

Equipment

1. Falling head permeameter
2. Balance, sensitive to 0.1 g
3. Thermometer
4. Stop watch

Falling Head Permeameter

A schematic diagram of a falling head permeameter is shown in Fig. 10–1. This consists of a specimen tube essentially the same as that used in the constant head test. The top of the specimen tube is connected to a burette by plastic tubing. The specimen tube and the burette are held vertically by clamps from a stand. The bottom of the specimen tube is connected to a plastic funnel by a plastic tube. The funnel is held vertically by a clamp from another stand. A scale is also fixed vertically to this stand.

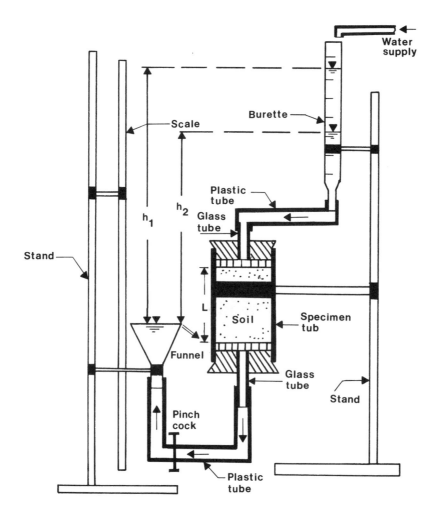

Figure 10–1. Schematic diagram of variable head permeability test setup.

Procedure

Steps 1 through 9: Follow the same procedure (i.e., Steps 1 through 9) as described in Chapter 9 for the preparation of the specimen.

10. Supply water, using a plastic tube from the water inlet to the burette. The water will flow from the burette to the specimen and then to the funnel. Check to see that there is no leak. Remove all air bubbles.

11. Allow the water to flow for some time in order to saturate the specimen. When the funnel is full, water will flow out of it into the sink.

12. Using the pinch cock, close the flow of water through the specimen. The pinch cock is located on the plastic pipe connecting the bottom of the specimen to the funnel.

13. Measure the head difference, h_1 (cm) (see Fig. 10–1).

    Note: Do not add any more water to the burette.

14. Open the pinch cock. Water will flow through the burette to the specimen and then out of the funnel. Record time (t) with a stop watch until the head difference is equal to h_2, in cm (Fig. 10–1). Close the flow of water through the specimen, using the pinch cock.

15. Determine the volume (V) of water that is drained from the burette in cm^3.

16. Add more water to the burette to make another run. Repeat Steps 13, 14, and 15. However, h_1 and h_2 should be changed for each run.

17. Record the temperature, T, of the water in °C.

Calculation

The coefficient of permeability can be expressed by the relation

$$k = 2.303 \, \frac{aL}{At} \, \log \frac{h_1}{h_2} \tag{10.1}$$

where a = inside cross-sectional area of the burette

[For an example for derivation, see Das (1990) under "References" at the back of this book.]

$$a = \frac{V}{(h_1 - h_2)} \tag{10.2}$$

Therefore

$$k = \frac{2.303 VL}{(h_1 - h_2) tA} \, \log \frac{h_1}{h_2} \tag{10.3}$$

where A = area of the specimen

As in Chapter 9,

$$k_{20°C} = k_{T°C} \, \frac{\eta_T}{\eta_{20}} \tag{10.4}$$

A sample calculation is shown in Table 10–1.

FALLING HEAD PERMEABILITY TEST

Description of soil *Uniform sand*

Sample No. _____ Location _____

Length of specimen, L _13.2_ (cm) Area of specimen, A _31.65_ (cm²)

Dry weight of specimen _726.9_ (g) G_s _2.66_ Void ratio, e _____

Test No.	h_1 (cm)	h_2 (cm)	Test duration, t (sec)	Temperature of water, T (°C)	Volume of water, V (cm³)	$k = \dfrac{2.303\,VL}{(h_1 - h_2)\,tA} \log \dfrac{h_1}{h_2}$
1	85.0	24.0	15.4	25	64	0.036
2	75.0	20.0	15.3	25	58	0.038
3	65.0	20.0	14.4	25	47	0.036

$k_{20°C}$ = _0.037(0.89)_ = _0.033_   Average k = _0.037_ cm/sec

Table 10–1

11

Standard Proctor Compaction Test

Introduction

For construction of highways, airports and other structures, compaction of soils is necessary to improve its strength. Proctor (1933) developed a laboratory compaction test procedure to determine the maximum dry unit weight of compaction for soils which can be used for specification of field compaction. This test is referred to as the *standard Proctor compaction test* and will be described in this chapter.

Equipment

1. Compaction mold
2. No. 4 sieve
3. Standard Proctor hammer [5.5 lb (24.5 N)]
4. Balance, sensitive up to 0.01 lb
5. Balance, sensitive up to 0.1 g
6. Large flat pan
7. Jack
8. Steel straight edge
9. Moisture cans
10. Drying oven
11. Plastic squeeze bottle with water

Fig. 11–1 shows the necessary equipment required for the compaction test (except the jack, the balances, and the oven).

Proctor Compaction Mold and Hammer

A schematic diagram of the Proctor compaction mold, which is 4 in. (101.6 mm) in diameter and 4.584 in. (116.4 mm) in height, is shown in Fig. 11–2a. There is a base plate

Figure 11–1. Equipment for Proctor compaction test.

and an extension that can be attached to the top and bottom of the mold, respectively. The inside volume of the mold is $\frac{1}{30}$ ft^3 (943.9 cm^3).

Fig. 11–2b shows the schematic diagram of a standard Proctor hammer. The hammer can be lifted and dropped through a vertical distance of 12 in. (304.8 mm).

Procedure

1. Obtain about ~~10 lb (44.48 N)~~ of air-dry soil on which the compaction test is to be conducted. Break all the soil lumps.

2. ~~Sieve the soil on a No. 4 sieve.~~ Collect all of the minus–4 material in a large pan. This should be about 6 lb (26.69 N) or more.

3. Add enough water to the minus–4 material, and mix it in thoroughly, to bring the moisture content up to about 5%.

4. Determine the weight of the Proctor mold + base plate (*not the extension*), W_1, in lb.

5. Now attach the extension to the top of the mold.

6. Pour the moist soil into the mold in *three* equal layers. Each layer should be compacted uniformly by the *standard Proctor hammer 25 times* before the next layer of loose soil is poured into the mold.

   Note: The layers of loose soil that are being poured into the mold should be such that, *at the end of the three-layer compaction*, the soil should extend *slightly above* the top of the rim of the compaction mold.

$$\omega = \frac{m_w}{m_s}$$

$$12 = \frac{m_w}{370.4}$$

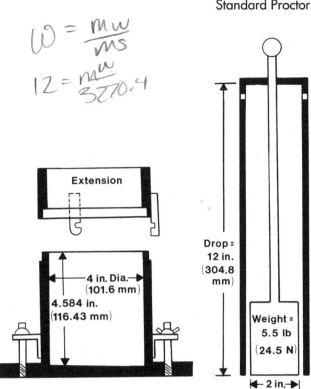

Figure 11-2. Standard Proctor mold and hammer.

7. Remove the top attachment from the mold. Be careful not to break off some of the compacted soil inside the mold while removing the top attachment.

8. Using a straight edge, trim the excess soil above the mold (Fig. 11–3). Now the top of the compacted soil will be even with the top of the mold.

9. Determine the weight of the mold + base plate + compacted moist soil in the mold (W_2, in lb).

10. Remove the base plate from the mold. Using a jack, extrude the compacted soil cylinder from the mold.

11. From the moist soil extruded in Step 10, collect a moisture sample in a moisture can and put it into the oven to dry to a constant weight.

12. Break the rest of the compacted soil (to No. 4 size) by hand, and mix it with the leftover moist soil in the pan. Add more water, and mix it, to raise the moisture content by *about 2%*.

13. Repeat Steps 6 through 12. In this process, the weight of the mold + base plate + moist soil (W_2) will first increase with the increase of the moisture content and then decrease. Continue the test until at least two successive down readings are obtained.

14. The next day, determine the dry weight of soils in the moisture cans (Step 11).

Figure 11–3. Excess soil being trimmed (Step 8).

Calculation

1. Moist unit weight for a given test =

$$\gamma = \frac{\text{weight of moist soil (lb)}}{\text{volume of mold } (\text{ft}^3)} = \frac{W_2 - W_1}{\dfrac{1}{30}}$$

$$= 30\,(W_2 - W_1)\,\text{lb}/\text{ft}^3 \qquad (11.1)$$

2. Dry unit weight for a given test =

$$\gamma_d = \frac{\gamma}{1 + \dfrac{w\,(\%)}{100}}$$

where w = moisture content

Graph

Plot a graph of γ_d vs. $w(\%)$. Join these points by a smooth curve. The ordinate of the peak point of the curve will give the *maximum dry unit weight of compaction*, and the corresponding moisture content will be the *optimum moisture content*.

A sample calculation for a Proctor test is given in Table 11–1, and the corresponding graph of γ_d vs. $w(\%)$ is shown in Fig. 11–4.

PROCTOR COMPACTION TEST

Description of soil _Light brown clayey silt_

Sample No. _2_   Location _Pine Street_

Volume of mold _$^1/30$_   Weight of hammer _5.5 lb_

No. of blows/layer _25_   No. of layers _3_ G_s _2.68_

Test No.	Weight of mold, W_1 (lb)	Weight of mold + moist soil, W_2 (lb)	Weight of moist soil, $W_2 - W_1$ (lb)	Moist unit weight, $30(W_2 - W_1)$ (lb/ft^3)	Moisture[a] content, w (%)	Dry unit weight, γ_d (lb/ft^3)
1	10.348	14.185	3.837	115.11	8.74	105.86
2		14.407	4.059	121.77	10.27	110.43
3		14.526	4.178	125.34	10.93	112.99
4		14.629	4.281	128.43	12.52	114.14
5		14.509	4.161	124.83	15.03	108.52
6		14.466	4.118	123.54	18.70	103.99

[a] To be obtained from the following table.

Moisture Content Determination

Test No.	1	2	3	4	5	6
Can No.	202	212	222	242	206	504
Weight of can (g)	54.00	53.25	53.25	54.00	54.75	40.8
Weight of can + moist soil (g)	253.00	354.01	439.00	490.00	422.83	243.00
Weight of can + dry soil (g)	237.00	326.00	401.01	441.50	374.72	211.1
Moisture content (%)	8.74	10.27	10.93	12.52	15.03	18.79

Table 11-1

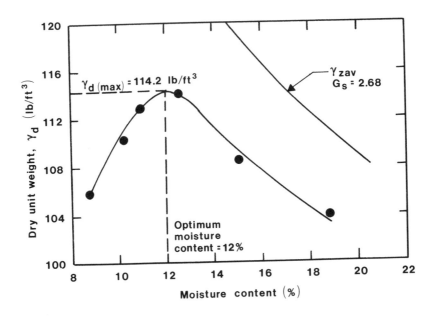

Figure 11–4. Plot at γ_d vs. w (%) for test results reported in Table 11–1.

Zero-Air-Void Curve

The maximum theoretical dry unit weight of compaction at a given moisture content will occur when there are no air voids left in the compacted soil. This can be given by

$$\gamma_{d\,(\text{theory}-\text{max})} = \gamma_{zav} = \frac{\gamma_w}{\dfrac{w(\%)}{100} + \dfrac{1}{G_s}}$$

where γ_{zav} = zero-air-void unit weight
γ_w = unit weight of water
w = moisture content
G_s = specific gravity of soil solids.

Since the values of γ_w and G_s will be known, several values of $w\%$ can be assumed and γ_{zav} can be calculated. These values are plotted on the same graph showing the experimental values of γ_d vs. $w\%$. As an example, the γ_{zav} plot for the soil reported in Table 11–1 is plotted in Fig. 11–4.

Note: For a given soil, *no portion* of the experimental curve of γ_d vs. $w\%$ should plot to the *right* of the zero-air-void curve.

General Comments

In most of the specifications for earth work, it is required to achieve a compacted field dry unit weight of 90 to 95% of the maximum dry unit weight obtained in the laboratory. This is sometimes referred to as relative compaction, R, or

$$R\,(\%) = \frac{\gamma_{d\,(\text{field})}}{\gamma_{d\,(\text{max} - \text{lab})}} \times 100 \tag{11.3}$$

For granular soils, it can be shown that

$$R\,(\%) = \frac{R_o}{1 - D_r\,(1 - R_o)} \times 100 \tag{11.4}$$

where D_r = relative density of compaction

$$R_o = \frac{\gamma_{d\,(\text{max})}}{\gamma_{d\,(\text{min})}} \tag{11.5}$$

Compaction of cohesive soil will influence its structure, coefficient of permeability, one-dimensional compressibility, and strength. For further discussion on this, refer to Das (1990).

12
Modified Proctor Compaction Test

Introduction

In the preceding chapter, we have seen that water generally acts as a lubricant between solid particles during the compaction process of soils. In the initial stages of compaction, the dry unit weight of compaction increases due to this reason. However, another factor that will control the dry unit weight of compaction of a soil at a given moisture content is the energy of compaction. For the standard Proctor compaction test, the energy of compaction can be given by

$$\frac{(3 \text{ layers})\,(25 \text{ blows/layer})\,(5.5 \text{ lb})\,(1 \text{ ft/blow})}{{}^1\!/_{30}\ \text{ft}^3}$$

$$= 12,375 \frac{\text{ft} \cdot \text{lb}}{\text{ft}^3}\,(593 \text{ kJ/m}^3)$$

The modified Proctor compaction test is a standard test procedure for compaction of soil using a higher energy of compaction. In this test, the compaction energy is equal to

$$56,250 \frac{\text{ft} \cdot \text{lb}}{\text{ft}^3}\,(2694 \text{ kJ/m}^3)$$

Equipment

The equipment required for the modified Proctor compaction test is the same as in Chapter 11, with the exception of Item 3. The hammer used for this test weighs 10 lb

Figure 12–1. Comparison of the standard and modified
Proctor compaction hammer. *Note:* The
left-side hammer is for the modified
Proctor compaction test.

(44.48 N) and drops through a vertical distance of 18 in. (457.2 mm). Fig. 12–1 shows
the standard and modified Proctor test hammers side by side.

The compaction mold used in this test is the same as described in Chapter 11 [i.e.,
volume = $\frac{1}{30}$ ft^3 (943.9 cm^3)].

Procedure
The procedure is the same as described in Chapter 11, except for Item 6. The moist soil
has to be poured into the mold in five equal layers. Each layer has to be compacted by
the modified Proctor hammer with 25 blows per layer.

Calculation, Graph, and Zero-Air-Void Curve
Same as in Chapter 11.

General Comments

1. The modified Proctor compaction test results for the same soil as reported in Table 11–1 and Fig. 11–4 are shown in Fig. 12–2. A comparison of γ_d vs. $w(\%)$ curves obtained from standard and modified Proctor tests show that
   (a) The maximum dry unit weight of compaction increases with the increase of the compacting energy, and
   (b) The optimum moisture content decreases with the increase of the energy of compaction.

2. In Chapters 11 and 12, the laboratory test outlines given for compaction tests use the following:
   Volume of mold = $\frac{1}{30}$ ft^3
   Number of blows = 25

These values are generally used for fine-grained soils that pass through No. 4 sieve. However, ASTM and AASHTO have four different suggested methods for each test (i.e., the standard Proctor compaction test and the modified Proctor compaction test) that reflect the size of the mold, the number of blows per layer, and the maximum particle size in a soil used for testing. Summaries of these methods are given in Tables 12–1 and 12–2.

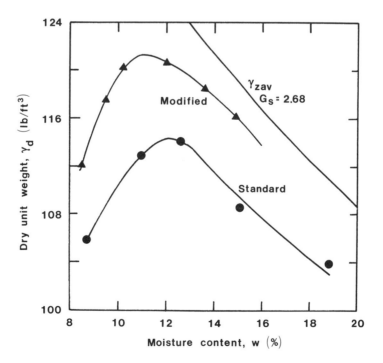

Figure 12–2. Comparison of standard and modified Proctor compaction test results for the soil reported in Table 11–1.

Table 12–1. Summary of Standard Proctor Compaction Test Specifications (ASTM D-698, AASHTO T-99)

Description	Method A	Method B	Method C	Method D
Mold:				
Volume (ft^3)	$\frac{1}{30}$	$\frac{1}{13.33}$	$\frac{1}{30}$	$\frac{1}{13.33}$
Height (in.)	4.58	4.58	4.58	4.58
Diameter (in.)	4	6	4	6
Weight of hammer (lb)	5.5	5.5	5.5	5.5
Height of drop of hammer (in.)	12	12	12	12
Number of layers of soil	3	3	3	3
Number of blows per layer	25	56	25	56
Test on soil fraction passing sieve	No. 4	No. 4	$\frac{3}{4}$ in.	$\frac{3}{4}$ in.

Table 12–2. Summary of Modified Proctor Compaction Test Specifications (ASTM D-1557, AASHTO T-180)

Description	Method A	Method B	Method C	Method D
Mold:				
Volume (ft^3)	$\frac{1}{30}$	$\frac{1}{13.33}$	$\frac{1}{30}$	$\frac{1}{13.33}$
Height (in.)	4.58	4.58	4.58	4.58
Diameter (in.)	4	6	4	6
Weight of hammer (lb)	10	10	10	10
Height of drop of hammer (in.)	18	18	18	18
Number of layers of soil	5	5	5	5
Number of blows per layer	25	56	25	56
Test on soil fraction passing sieve	No. 4	No. 4	$\frac{3}{4}$ in.	$\frac{3}{4}$ in.

13

Determination of Field Unit Weight of Compaction by Sand Cone Method

Introduction

In the field, during earth compaction work, it is necessary from time to time to check the compacted dry unit weight of soil and compare it with the specification drawn up for the construction. One of the simplest methods of determining the field unit weight of compaction is by the sand cone method, which will be described in this chapter.

Equipment

1. Sand cone apparatus. This consists of a one-gallon glass or plastic bottle with a metal cone attached to it.
2. Base plate
3. One-gallon can with cap
4. Tools to dig a small hole in the field
5. Balance
6. 20-30 Ottawa sand
7. Proctor compaction mold without attached extension
8. Steel straight edge

Figure 13–1 shows the assembly of the equipment necessary for determination of the field unit weight.

Figure 13–1. Assembly of equipment necessary for the determination of field unit weight of compaction.

Procedure — Laboratory Work

1. Determine the dry unit weight of the 20-30 Ottawa sand that will be used in the field. This can be done by taking a Proctor compaction mold and, using a spoon, filling it with Ottawa sand. Avoid any vibration or other means of compaction of the sand poured into the mold. When the mold is full, strike off the top of the mold with the steel straight edge. Determine the weight of the sand in the mold (W_1). The dry unit weight of the Ottawa sand can then be given as

$$\gamma_{d\,(\text{sand})} = \frac{W_1}{V_1} \qquad (13.1)$$

where $\gamma_{d\,(\text{sand})}$ = dry unit weight of Ottawa sand
  V_1 = volume of Proctor mold = $\frac{1}{30}$ ft^3

2. Calibrate the cone. That is, we need to determine the weight of the Ottawa sand that is required to fill the cone. This can be done as follows: Fill the one-gallon bottle with Ottawa sand. Determine the weight of the bottle + cone + sand (W_2). Close the valve of the cone which is attached to the bottle. Place the base plate on a flat surface. Turn the bottle with the cone attached to it upside down, and place the open mouth of the cone in the center hole of the base plate (Fig. 13–2). Open the cone valve. Sand will flow out of the bottle and gradually fill the cone. When the cone is filled with sand, the flow of sand from the bottle will stop.

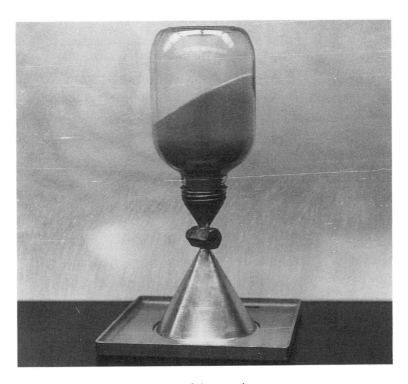

Figure 13–2. Calibration of the sand cone.

Close the cone valve. Remove the bottle and cone combination from the base plate, and determine its weight (W_3). The weight of sand necessary to fill the cone can now be determined as

$$W_c = W_2 - W_3$$

3. Determine the weight of the gallon can without the cap (W_4).
4. Before proceeding to the field, fill the one-gallon bottle (with the sand cone attached to it) with sand. Close the valve of the cone. Determine the weight of the bottle + cone + sand (W_5).

Procedure — Field Work

5. Now proceed to the field with the bottle with the cone attached to it (filled with Ottawa sand — Step 4), the base plate, the digging tools, and the one-gallon can with its cap.
6. Place the base plate on a level ground in the field. Under the center hole of the base plate, dig a hole in the ground using the digging tools. The volume of the hole should be smaller than the volume of the sand in the bottle minus the volume of the cone.
7. Remove *all* the loose soil from the hole, and put it in the gallon can. Close the cap tightly so as not to lose any moisture. Be careful not to move the base plate.

8. Turn the gallon bottle filled with sand with cone attached to it upside down, and place it on the center of the base plate. Open the valve of the cone. Sand will flow from the bottle to fill the hole in the ground and the cone. When the flow of sand from the bottle stops, close the valve of the cone and remove it.

9. Bring all the equipment back to the laboratory. Determine the weight (W_6) of the gallon can + moist soil from the field (without the cap). Also, determine the weight of the bottle + can + sand after use (W_7).

10. Put the gallon can with the moist soil in the oven to dry to a constant weight. Determine the weight of the can + oven-dry soil (W_8).

Calculation

1. Determine the moisture content of the soil in the field as

$$w\% = \frac{W_6 - W_8}{W_8 - W_4}(100) \qquad (13.3)$$

2. Determine the moist unit weight of the soil in the field as follows

$$\gamma = \frac{\text{weight of moist soil from the hole}}{\text{volume of the hole}} \qquad (13.4)$$

$$\text{Weight of moist soil from the hole} = W_6 - W_4 \qquad (13.5)$$

$$\text{Volume of the hole} = \frac{W_5 - W_7 - W_c}{\gamma_{d\,(\text{sand})}} \qquad (13.6)$$

3. Determine the dry unit weight of soil in the field as

$$\gamma_{d\,(\text{field})} = \frac{\gamma}{1 + \dfrac{w\,(\%)}{100}} \qquad (13.7)$$

A sample calculation is shown in Table 13–1.

General Comments

There are at least two other methods for determination of the field unit weight of compaction. They are the *rubber balloon method* (ASTM D-2167) and the *use of the nuclear density meter*. The procedure for the rubber balloon method is similar to the sand cone method, in that a test hole is made and the moist weight of the soil removed from the hole and its moisture content are determined. However, the volume of the hole is determined by introducing into it a rubber balloon filled with water from a calibrated vessel, from which the volume can be read directly.

Nuclear density meters are now used in several large projects to determine the compacted dry unit weight of a soil. The density meters operate either in drilled holes or from the ground surface. The instrument measures the weight of the wet soil per unit volume and also the weight of water present in a unit volume of soil. The dry unit weight of compacted soil can be determined by subtracting the weight of water from the moist unit weight of soil.

FIELD UNIT WEIGHT—SAND CONE METHOD

Calibration of Unit Weight of Ottawa Sand

Weight of sand in the mold, W_1	3.31 lb
Volume of mold, V_1	1/30 ft^3
$\gamma_{d\,(\text{sand})} = \dfrac{W_1}{V_1}$	99.3 lb/ft^3

Calibration Cone

Weight of bottle + cone + sand (before use), W_2	15.17 lb
Weight of bottle + cone + sand (after use), W_3	14.09 lb
Weight of sand to fill the cone, $W_c = W_2 - W_3$	1.08 lb

Results from Field Tests

Weight of bottle + cone + sand (before use), W_5	15.42 lb
Weight of bottle + cone + sand (after use), W_7	11.74 lb
Volume of hole, $\dfrac{W_5 - W_7 - W_c}{\gamma_{d\,(\text{sand})}}$	0.0262 ft^3
Weight of gallon can, W_4	0.82 lb
Weight of can + moist soil, W_6	3.92 lb
Weight of can + dry soil, W_8	3.65 lb
Weight of moist soil, $W_6 - W_4$	3.1 lb
Moist unit weight of soil in field, $\gamma = \dfrac{(W_6 - W_4)\,\gamma_{d\,(\text{sand})}}{W_5 - W_7 - W_c}$	$\dfrac{3.1}{0.0262} = 118.32$ lb/ft^3
Moisture content in the field, $w\,(\%) = \dfrac{W_6 - W_8}{W_8 - W_4}(100)$	9.44%
Dry unit weight in the field, $\gamma_d = \dfrac{\gamma}{1 + \dfrac{w\%}{100}}$	108.11 lb/ft^3

Table 13-1

14

Direct Shear Test on Sand

Introduction

The shear strength of a sand can be expressed by the equation

$$s = \sigma' \tan\phi \qquad (14.1)$$

where s = shear strength
σ' = effective normal stress
ϕ = angle of friction of soil.

The angle of friction, ϕ, is a function of the relative density of compaction of sand, grain size, shape, and distribution in a given soil mass. The general range of the angle of friction of sand with relative density is shown in Fig. 14–1.

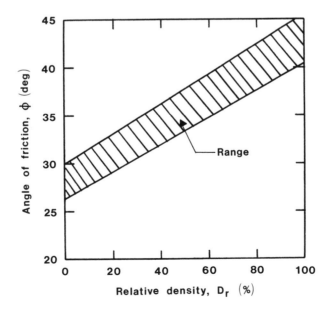

Figure 14–1. Range of the variation of the angle of friction of sand with relative density of compaction.

Figure 14–2. A direct shear test machine.

Equipment

1. Direct shear test machine (strain-controlled)
2. Balance, sensitive to 0.1 g
3. Large porcelain evaporating dish
4. Tamper (for compacting sand in the direct shear box)
5. Spoon

Fig. 14–2 shows a direct shear test machine. It mainly consists of a direct shear box, which is split into two halves (i.e., top and bottom) and holds the soil specimen, a proving ring to measure the horizontal load applied to a specimen; two dial gauges (one horizontal and one vertical) to measure the deformation of the soil during the test; and a yoke by which a vertical load can be applied to the soil specimen. A horizontal load to the top half of the shear box is applied by a motor and gear arrangement. In a strain-controlled unit, the rate of movement of the top half of the shear box can be controlled.

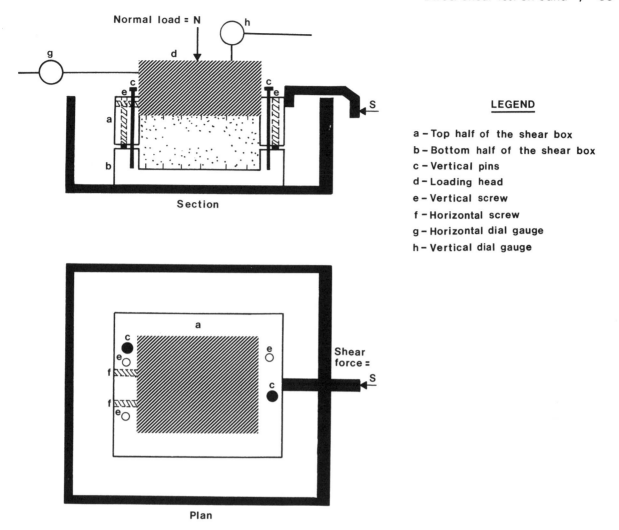

Normal load = N

LEGEND

a – Top half of the shear box
b – Bottom half of the shear box
c – Vertical pins
d – Loading head
e – Vertical screw
f – Horizontal screw
g – Horizontal dial gauge
h – Vertical dial gauge

Section

Shear force = S

Plan

Figure 14-3. Schematic diagram of a direct shear testing box.

Fig. 14–3 shows the schematic diagram of the shear box. The shear box is split into two halves—top and bottom. The top and bottom halves of the shear box can be held together by two vertical pins. There is a loading head that can be slipped from the top of the shear box to rest on the soil specimen inside the box. There are also three vertical screws and two horizontal screws on the top half of the shear box.

Procedure

1. Remove the shear box assembly. Back off the three vertical and two horizontal screws. Remove the loading head. Insert the two vertical pins to keep the two halves of the shear box together.

2. Weigh some dry sand in a large porcelain dish. Fill the shear box with sand in small layers. A tamper may be used to compact the sand layers. The top of the compacted specimen should be about $\frac{1}{4}$ in. (6.4 mm) below the top of the shear box. Level the surface of the sand specimen. Now determine the weight of the sand left in the porcelain dish. The difference between the initial and final weights of sand is the weight of the sand in the shear box (W).

3. Determine the dimensions of the soil specimen (i.e., length, width, and height of the specimen).

4. Slip the loading head down from the top of the shear box to rest on the soil specimen.

5. Put the shear box assembly in place in the direct shear machine.

6. Apply the desired normal load, N, on the specimen. This can be done by hanging dead weights to the vertical load yoke. The top crossbars will rest on the loading head of the specimen, which, in turn, rests on the soil specimen.

   Note: In the equipment shown in Fig. 14–2, the weights of the hanger, the loading head, and the top half of the shear box can be tared. In some other equipment, if taring is not possible, the normal load should be calculated as N = load hanger + weight of yoke + weight of loading head + weight of top half of the shear box.)

7. Remove the two vertical pins (which were inserted in Step 1 to keep the two halves of the shear box together).

8. Advance the three vertical screws that are located on the side walls of the top half of the shear box. This is done to separate the two halves of the box. The space between the two halves of the box should be slightly larger than the largest grain size of the soil specimen (by visual observation).

9. Set the loading head by tightening the two horizontal screws located on the top half of the shear box. Now back off the three vertical screws. After doing this there will be no connection between the two halves of the shear box except the soil.

10. Attach the horizontal and vertical dial gauges (0.001 in./small div) to the shear box to measure the displacements during the test.

11. Apply horizontal load, S, to the top half of the shear box. The rate of shear displacement should be between 0.1 to 0.02 in./min (2.54 to 0.51 mm/min). For every tenth small division displacement in the horizontal dial gauge, record the readings of the vertical dial gauge and the proving ring dial gauge (which measures horizontal load, S). Continue this until after
    (a) the proving ring dial gauge reading reaches a maximum and then falls, or
    (b) the proving ring dial gauge reading reaches a maximum and then remains constant.

12. Repeat the test (Steps 1 to 11) at least two more times. For each test, the dry unit weight of compaction of sand specimen should be the same as that of the first specimen (Step 2).

Calculation

A sample calculation of a direct shear test in sand for one normal load (N) is shown in Table 14–1. Referring to Table 14–1, the calculations can be done as follows:

1. Determine the dry unit weight of specimen, γ_d

$$\gamma_d = \frac{W}{LBH} \qquad (14.2)$$

where W = weight of the specimen
$L, B,$ and H = length, width, and height of the specimen.

DIRECT SHEAR TEST ON SAND

Description of soil _Uniform sand_

Sample No. _2_   Location _Cedar Court_

Size of specimen _2 in. x 2 in. x 1.31 in. (Height)_

Weight of specimen, W _143.1 g_   Dry unit weight _104 lb/ft^3_

G_s _2.66_   Void ratio, e _0.596_   Normal load, N _56 lb_

Normal stress, σ' _14 lb/in.^{2}_

Proving ring calibration factor _0.31 lb/div._

Horizontal displacement (in.) (1)	Vertical displacement* (in.) (2)	No. of div. in proving ring dial gauge (3)	Shear force, S (lb) (4)	Shear stress, τ (lb/in.^{2}) (5)
0	0	0	0	0
0.01	+ 0.001	45	13.95	3.49
0.02	+ 0.002	76	23.56	5.89
0.03	+ 0.004	95	29.76	7.44
0.04	+ 0.006	112	34.72	8.68
0.05	+ 0.008	124	38.44	9.61
0.06	+ 0.009	129	39.99	10.00
0.07	+ 0.010	125	38.75	9.69
0.08	+ 0.010	119	36.89	9.22
0.09	+ 0.009	114	35.34	8.84
0.10	+ 0.008	109	33.79	8.45
0.11	+ 0.008	108	33.48	8.37
0.12	+ 0.008	105	32.55	8.14

* + sign means expansion

Table 14–1

2. Determine the void ratio of the specimen, e.

$$e = \frac{G_s \gamma_w}{\gamma_d} - 1 \qquad (14.3)$$

where G_s = specific gravity of soil solids
γ_w = unit weight of water.

3. Determine the normal effective stress on the specimen, σ'.

$$\sigma' = \frac{N}{L \times B} \qquad (14.4)$$

4. The horizontal, vertical, and proving ring dial gauge readings are obtained from the test (Columns 1, 2, and 3 in Table 14–1).

5. For any given set of horizontal and vertical dial gauge readings, calculate the shear force.

S = (No. of divisions in proving ring dial gauge, i.e., Col. 3) × (14.5)
(proving ring calibration factor)

6. Calculate the shear stress as

$$\tau = \frac{\text{shear force}}{\text{area of the specimen}} = \frac{S}{L \times B} \qquad (14.6)$$

Note: A separate data sheet has to be used for each test (i.e., for each normal stress, σ').

Graph

1. For each normal stress, plot a graph of τ (Col. 5) vs. horizontal displacement (Col. 1) (as shown in Fig. 14–4 for the results obtained from Table 14–1). On the bottom of the same graph paper, using the same horizontal scale, plot a graph of vertical displacement (Col. 2) vs. horizontal displacement (Col. 1). There will be at least three such plottings (one for each value of σ'). Determine the shear stresses at failure, s, from each τ vs. horizontal displacement graph (as shown in Fig. 14–4).

2. Plot a graph of shear strength, s, vs. normal stress, σ'. This graph will be a straight line passing through the origin. Fig. 14–5 shows such a plot for the sand reported in Table 14–1. The angle of friction of the soil can be determined from the slope of the straight line plot of s vs. σ' as

$$\phi = \tan^{-1}\left(\frac{s}{\sigma'}\right) \qquad (14.7)$$

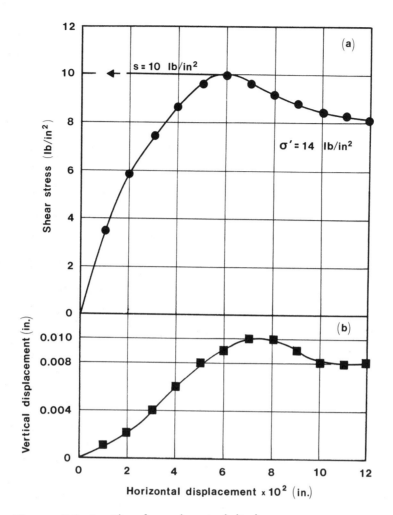

Figure 14-4. Plot of τ and vertical displacement vs. horizontal displacement for the direct shear test reported in Table 14-1.

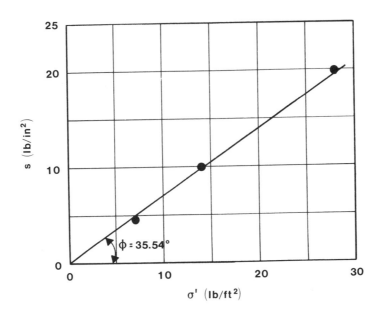

Figure 14–5. Plot of *s* vs. σ′ for the sand reported in Table 14–1.
Note: The results for tests with σ′ = 7 lb/in.² and 28 lb/in.² are not shown in Table 14–1.

General Comments

Typical values of the drained angle of friction, φ, for sands are given below:

Soil Type	φ (deg)
Sand: Round-grained	
Loose	28–32
Medium	30–35
Dense	34–38
Sand: Angular-grained	
Loose	30–36
Medium	34–40
Dense	40–45

15

Unconfined Compression Test

Introduction

Shear strength of a soil can be given by Coulomb's equation as

$$s = c + \sigma \tan \phi \qquad (15.1)$$

where s = shear stress
  c = cohesion
  σ = normal stress
  ϕ = angle of friction.

For *undrained tests of saturated clayey soils* ($\phi = 0$ condition).

$$s = c_u \qquad (15.2)$$

where c_u = undrained cohesion (or undrained shear strength).

The unconfined compression test is a quick method of determining the value of c_u for a clayey soil. The unconfined strength is given by the relation [for further discussion, see any soil mechanics text, e.g., Das (1990)].

$$q_u = \frac{c_u}{2} \qquad (15.3)$$

where q_u = unconfined compression strength.

The unconfined compression strength is determined by applying an axial stress to a cylindrical soil specimen with no confining pressure and observing the axial strains corresponding to various stress levels. The stress at which failure in the soil specimen occurs is referred to as the unconfined compression strength (Fig. 15–1). For *saturated* clay specimens, the unconfined compression strength decreases with the increase of moisture content

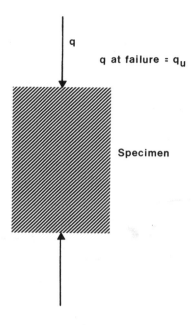

q

q at failure = q_u

Specimen

Figure 15–1.

of the soil. For *unsaturated* soils, with the dry unit weight remaining constant, the unconfined compression strength decreases with the increase of the degree of saturation.

Equipment

1. Unconfined compression testing device
2. Specimen trimmer and accessories (if undisturbed field specimen is used)
3. Harvard miniature compaction device and accessories (if a specimen is to be molded for classroom work)
4. Scale
5. Balance, sensitive to 0.01 g
6. Oven
7. Porcelain evaporating dish

Unconfined Compression Test Machine

An unconfined compression test machine in which *strain-controlled* tests can be performed is shown in Fig. 15–2. The machine essentially consists of a top and a bottom loading plate. The bottom of a proving ring is attached to the top loading plate. The top of the proving ring is attached to a cross-bar, which, in turn, is fixed to two metal posts. The bottom loading plate can be moved up or down.

Figure 15–2. An unconfined compression testing machine.

Procedure

1. Obtain a soil specimen for the test. If it is an undisturbed specimen, it has to be trimmed to the proper size by using the specimen trimmer. For classroom laboratory work, specimens at various moisture contents can be prepared using a Harvard miniature compaction device.

   The cylindrical soil specimen should have a height-to-diameter (*L*-to-*D*) ratio of between 2 and 3. In many instances, specimens with diameters of 1.4 in. (35.56 mm) and heights of 3.5 in. (88.9 mm) are used.

2. Measure the diameter and height of the specimen, and take the moist weight of the specimen.

3. Place the specimen centrally between the two loading plates of the unconfined compression machine. Move the top loading plate very carefully just to touch the top of the specimen. Set the proving ring dial gauge to zero.

   A dial gauge [each small division of the gauge should be equal to 0.001 in. (0.0254 mm) of vertical travel] should be attached to the unconfined compression machine to record the vertical upward movement (i.e., compression of the specimen during testing) of the bottom loading plate. Set this dial gauge to zero.

4. Turn the machine on. Record loads (i.e., proving ring dial gauge readings) and the corresponding specimen deformations. During the load application, the rate of *vertical strain* should be adjusted to $\frac{1}{2}$% to 2% per minute. At the initial stage of the test, usually readings are taken every 0.01 in. (0.254 mm) of specimen deformation. However, this can be varied to every 0.02 in. (0.508 mm) of specimen

Figure 15–3. A soil specimen after failure.

deformation at a later stage of the test when the load-deformation curve begins to flatten out.

5. Continue taking readings until
    (a) Load reaches a peak and then decreases; or
    (b) Load reaches a maximum value and remains approximately constant there-after (take about 5 readings after it reaches the peak value); or
    (c) Deformation of the specimen is past 20% strain before reaching the peak. This may happen in the case of soft clays.
    Fig. 15–3 shows a soil specimen after failure.
6. Unload the specimen by lowering the bottom loading plate.
7. Remove the specimen from between the two loading plates.
8. Draw a free-hand sketch of the specimen after failure. Show the nature of the failure.
9. Put the specimen in a porcelain evaporating dish, and determine the moisture content after drying it in an oven to a constant weight.

Calculation

For each set of readings (refer to Table 15–1):

1. Calculate the vertical strain (Col. 2), ϵ.

$$\epsilon = \frac{\Delta L}{L} \tag{15.4}$$

where ΔL = total vertical deformation of the specimen
       L = original height of specimen.

UNCONFINED COMPRESSION TEST

Description of soil _Light brown clay_

Sample No. _3_   Location _Spruce Lane_

Moist weight of specimen _149.8 g_ Moisture content _12_ %

Length of specimen, L _3 in._   Diameter, D _1.43 in._   Area, A_0 _1.605 in.^{2}_

Proving ring calibration factor: 1 div. = _0.264 lb._

Specimen deformation = ΔL (in.) (1)	Vertical strain, $\epsilon = \dfrac{\Delta L}{L}$ (2)	Proving ring dial reading [No. of small divisions] (3)	Load = Col. 3 × calibration factor of proving ring (lb) (4)	Corrected area = $A_c = \dfrac{A_0}{1-\epsilon}$ (in.^{2}) (5)	Stress = $\dfrac{\text{Col. 4}}{\text{Col. 5}}$ (lb/in.^{2}) (6)
0	0	0	0	1.43	0
0.01	0.0033	12	3.168	1.435	2.208
0.02	0.0067	38	10.032	1.44	6.967
0.03	0.01	52	13.728	1.444	9.507
0.04	0.013	58	15.312	1.449	10.567
0.06	0.02	67	17.688	1.459	12.123
0.08	0.027	74	19.536	1.470	13.289
0.10	0.033	78	20.592	1.479	13.923
0.12	0.04	81	21.384	1.490	14.352
0.14	0.047	83	21.912	1.500	14.608
0.16	0.053	85	22.440	1.510	14.861
0.18	0.06	86	22.704	1.521	14.927
0.20	0.067	86	22.704	1.533	14.810
0.24	0.08	84	22.176	1.554	14.270
0.28	0.093	83	21.912	1.577	13.895
0.32	0.107	82	21.648	1.601	13.522
0.36	0.12	81	21.384	1.625	13.159

Table 15-1

2. Calculate the vertical load on the specimen (Col. 4).

$$\text{Load} = (\text{proving ring dial reading})(\text{calibration factor}) \qquad (15.5)$$

3. Calculate the corrected area of the specimen A_c (Col. 5).

$$A_c = \frac{A_0}{1 - \epsilon} \qquad (15.6)$$

where A_0 = initial area of cross-section of the specimen

$$= \frac{\pi}{4}D^2$$

4. Calculate the stress, σ, on the specimen (Col. 6).

$$\sigma = \frac{\text{Load}}{A_c} = \frac{\text{Col. 4}}{\text{Col. 5}} \qquad (15.7)$$

Graph

Plot the graph of stress (Col. 6) vs. strain, in percent (Col. 2 × 100). Determine the peak stress from this graph. This is the unconfined compression strength, q_u, of the specimen. *Note*: If 20% strain occurs before the peak stress, then the stress corresponding to 20% strain should be taken as q_u.

A sample calculation and graph are shown in Table 15–1 and Fig. 15–4.

General Comments

1. In the determination of unconfined compression strength, it is better to conduct tests on two to three identical specimens. The average value of q_u is the representative value.
2. Based on the value of q_u, the consistency of a cohesive soil is as follows:

Consistency	q_u (lb/ft^2)
Very soft	0–500
Soft	500–1000
Medium	1000–2000
Stiff	2000–4000
Very stiff	4000–8000

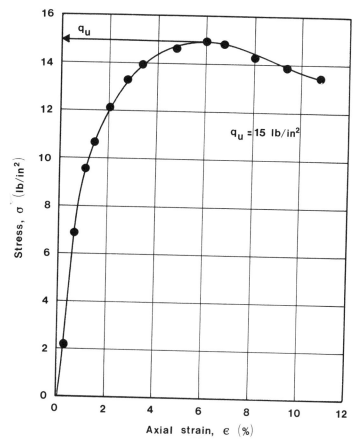

Figure 15–4. Plot of σ vs. ε (%) for the test results shown in Table 15–1.

3. For many naturally deposited clayey soils, the unconfined compression strength is greatly reduced when the soil is tested after remolding without any change of moisture content. This is referred to as *sensitivity* and can be defined as

$$S_t = \frac{q_{u\,(\text{undisturbed})}}{q_{u\,(\text{remolded})}} \tag{15.8}$$

The sensitivity of most clays ranges from 1 to 8. Based on the magnitude of S_t, clays can be described as follows:

Sensitivity, S_t	Description
1–2	Slightly sensitive
2–4	Medium sensitive
4–8	Very sensitive
8–16	Slightly quick
16–32	Medium quick
32–64	Very quick
> 64	Extra quick

16
Consolidation Test

Introduction

Consolidation is the process of time-dependent settlement of saturated clayey soil when subjected to an increased loading. In this chapter, the procedure for a one-dimensional laboratory consolidation test will be described, and methods of calculation to obtain the void ratio–pressure curve (e vs. $\log p$), the preconsolidation pressure, (p_c), and the coefficient of consolidation (c_v) will be outlined.

Equipment

1. Consolidation unit
2. Specimen trimming device
3. Wire saw
4. Balance, sensitive to 0.01 g
5. Stop watch
6. Moisture can
7. Oven

The consolidation unit consists of a consolidometer and a loading unit. The consolidometer can be either (*i*) a floating ring consolidometer (Fig. 16–1*a*) or (*ii*) a fixed ring consolidometer (Fig. 16–1*b*). The floating ring consolidometer usually consists of a brass ring in which the soil specimen is placed. One porous stone is placed at the top of the specimen and another porous stone at the bottom. The soil specimen in the ring with the two porous stones is placed on a base plate. A plastic ring surrounding the specimen fits into a groove on the base plate. Load is applied through a loading head that is placed on the top porous stone. In the floating ring consolidometer, compression of the soil specimen occurs from the top and bottom towards the center.

The fixed ring consolidometer consists essentially of the same components, i.e., a hollow base plate, two porous stones, a brass ring to hold the soil specimen, and a metal ring that can be fixed tightly to the top of the base plate. The ring surrounds the soil specimen. A stand pipe is attached to the side of the base plate. This can be used for

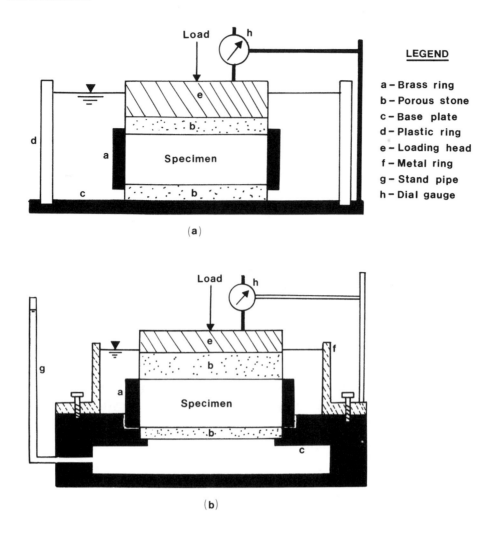

Figure 16–1. Schematic diagram of (a) floating ring consolidometer.
(b) fixed ring consolidometer.

permeability determination of soil. In the fixed ring consolidometer, the compression of the specimen occurs from the top towards the bottom.

The specifications for the loading devices of the consolidation unit vary depending on the manufacturer. Fig. 16–2 shows one type of loading device.

During the consolidation test, when load is applied to a soil specimen, the nature of variation of side friction between the surrounding brass ring and the specimen are different for the fixed ring and the floating ring consolidometer, and this is shown in Fig. 16–3 on page 100. In most cases, a side friction of 10% of the applied load is a reasonable estimate.

Procedure

1. Prepare a soil specimen for the test. The specimen is prepared by trimming an undisturbed natural sample obtained in shelby tubes. The shelby tube sample should be about $\frac{1}{4}$ in. to $\frac{1}{2}$ in. (6.35 mm to 12.7 mm) larger in diameter than the specimen diameter to be prepared for the test.

    Note: For classroom instructional purposes, a specimen can be molded in the laboratory.

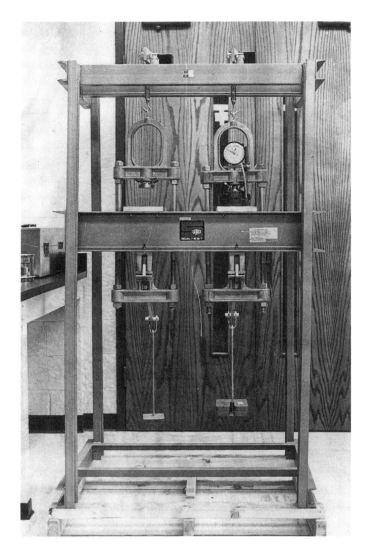

Figure 16–2. Consolidation loading assembly. In this assembly, two specimens can be simultaneously tested. Lever arm ratio for loading is 1:10.

2. Collect some excess soil that has been trimmed in a moisture can for moisture content determination.

3. Collect some of the excess soil trimmed in Step 1 for determination of the specific gravity of soil solids, G_s.

4. Determine the weight of the consolidation ring (W_1).

5. Place the soil specimen in the consolidation ring. Use the wire saw to trim the specimen flush with the top and bottom of the consolidation ring. Record the size of the specimen.

6. Determine the weight of the consolidation ring and the specimen (W_2).

7. Saturate the lower porous stone on the base of the consolidometer.

8. Place the soil specimen in the ring over the lower porous stone.

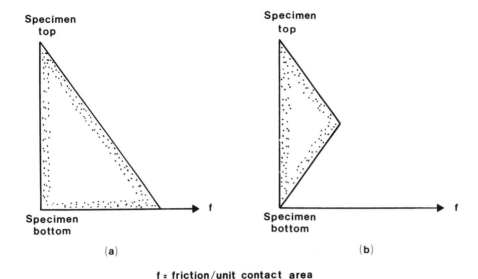

f = friction/unit contact area

Figure 16–3. Nature of variation of soil-ring friction per unit contact areas in (a) fixed ring consolidometer. (b) floating ring consolidometer.

9. Place the upper porous stone on the specimen in the ring.

10. Attach the top ring to the base of the consolidometer.

11. Add water to the consolidometer to submerge the soil and keep it saturated. In the case of the fixed ring consolidometer, the outside ring (attached to the top of the base) and the stand pipe connection attached to the base should be kept full with water. This needs to be done for the *entire period of the test.*

12. Place the consolidometer in the loading device.

13. Attach the vertical deflection dial gauge to measure the compression of soil. It should be fixed in such a way that the dial is at the beginning of its release run. The dial gauge used should be calibrated to read as 1 small div = 0.0001 in. (0.00254 mm).

14. Apply load to the specimen such that the magnitude of pressure, p, on the specimen is $\frac{1}{2}$ ton/ft^2 (47.88 kN/m^2). Take the vertical deflection dial gauge reading at the following times, t, counted from the time of the load application: 0 min., 0.25 min., 1 min., 2.25 min., 4 min., 6.25 min., 9 min., 12.25 min., 20.25 min., 25 min., 36 min., 60 min., 120 min., 240 min., 480 min., and 1440 min. (24 hr.).

15. The next day, add more load to the specimen such that the total magnitude of pressure on the specimen becomes 1 ton/ft^2 (95.76 kN/m^2). Take the vertical dial gauge readings at similar time intervals stated in Step 14. *Note:* Here we have $\frac{\Delta p}{p} = 1$. (where Δp = increase of pressure and p = the pressure before the increase).

16. Repeat Step 15 for soil pressure magnitudes of 2 ton/ft^2 (191.52 kN/m^2), 4 ton/ft^2 (383.04 kN/m^2), and 8 ton/ft^2 (766.08 kN/m^2), etc. *Note:* In all cases $\frac{\Delta p}{p} = 1$.

17. At the end of the test, remove the soil specimen and determine its moisture content.

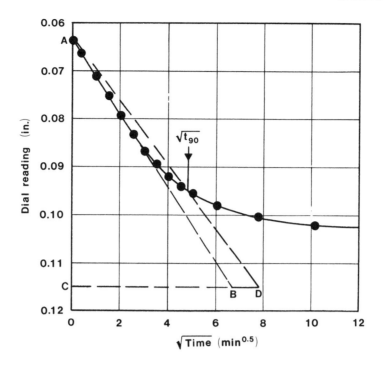

Figure 16–4. Plot of dial reading vs. $\sqrt{\text{time}}$ for the test results given in Table 16–1. Determination of t_{90} by square-root-of-time method.

Calculation and Graph

The calculation procedure for the test can be explained with reference to Tables 16–1 and 16–2 and Figs. 16–4, 16–5, and 16–6, which show the laboratory test results for a light brown clay.

1. Collect all the time vs. vertical dial readings data. Table 16–1 shows the results of a pressure increase from $p = 2$ ton/ft^2 to $p + \Delta p = 4$ ton/ft^2.

2. Determine the time for 90% primary consolidation, t_{90}, from each set of time vs. vertical dial readings. An example of this is shown in Fig. 16–4, which is a plot of the results of vertical dial reading vs. $\sqrt{\text{time}}$ given in Table 16–1. Draw a tangent AB to the initial consolidation curve. Measure the length BC. Plot the point D such that the length of $CD = 1.15$ times the length BC. Join AD. The abscissa of the point of intersection of the line AD with the consolidation curve will give $\sqrt{t_{90}}$. In Fig. 16–4, $\sqrt{t_{90}} = 4.75$ min.$^{1/2}$, so $t_{90} = (4.75)^2 = 22.56$ min. This technique is referred to as the *square-root-of-time-fitting method* (Taylor, 1942).

3. Determine the time for 50% primary consolidation, t_{50}, from each set of time vs. vertical dial readings. The procedure for this is shown in Fig. 16–5, which is a semilogarithmic plot (vertical dial reading in natural scale and time in log scale) for the set of readings shown in Table 16–1. Project the straight line portion of the primary consolidation downwards and the straight line portion of the secondary consolidation backwards. The point of intersection of these two lines is A. The vertical dial

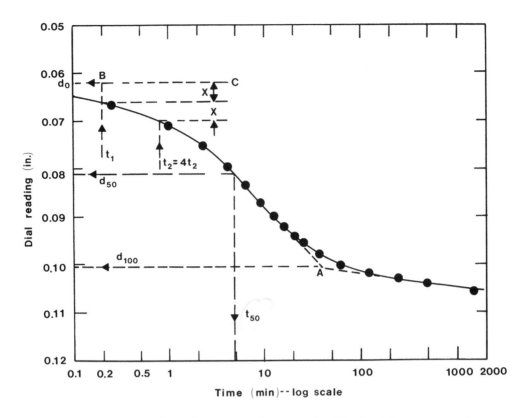

Figure 16–5. Logarithm-of-time curve fitting method for the laboratory results given in Table 16–1.

reading corresponding to point A is d_{100} (dial reading at 100% primary consolidation). Select times t_1 and $t_2 = 4t_1$. (*Note:* t_1 and t_2 should be within the top curved portion of the consolidation plot.) Determine the difference in dial readings, X, between times t_1 and t_2. Plot line BC, which is the vertical X distance above the point on the consolidation curve corresponding to time t_1. The vertical dial gauge reading corresponding to line BC is d_0, i.e., the reading for 0% consolidation. Determine the dial gauge reading corresponding to 50% primary consolidation as

$$d_{50} = \frac{d_0 + d_{100}}{2} \tag{16.1}$$

The time corresponding to d_{50} on the consolidation curve is t_{50}. This is the logarithm-of-time curve fitting method (Casagrande and Fadum, 1940). In Fig. 16–5, $t_{50} = 14.9$ min.

4. Complete the experimental data in Cols. 1, 2, 8, and 9 of Table 16–2. Columns 1 and 2 are obtained from time-dial reading tables (such as Table 16–1), and Cols. 8 and 9 are obtained from Steps 2 and 3, respectively.

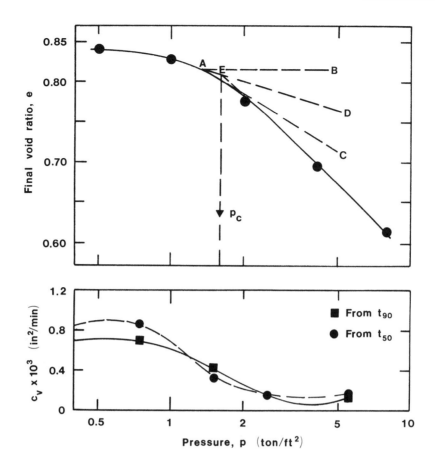

Figure 16–6. Plot of void ratio and the coefficient of consolidation against pressure for the soil reported in Table 16–2.

5. Determine the height of solids of the specimen in the mold as (see Table 16–2)

$$H_s = \frac{W_s}{\left(\dfrac{\pi}{4} D^2\right) G_s \gamma_w} \tag{16.2}$$

where H_s = height of solids
W_s = dry weight of soil specimen
D = diameter of the specimen
G_s = specific gravity soil solids
γ_w = unit weight of water.

6. In Table 16–2, determine the change in heights, ΔH, of the specimen due to load increments from p to $p + \Delta p$ (Col. 3). For example,

  $p = \frac{1}{2}$ ton/ft^2, final dial reading = 0.0283 in.
  $p + \Delta p = 1$ ton/ft^2, final dial reading = 0.0356 in.

Thus

  $\Delta H = 0.0356 - 0.0283 = 0.0073$ in.

CONSOLIDATION TEST
(Time vs. vertical dial reading)

Description of soil *Light brown clay*

Pressure on specimen *4 ton/ft²*

Clock time of load application *8:35 am*

Time after load application, t (min.)	$\sqrt{t}$ (min$^{0.5}$)	Vertical dial reading (in.)
0	0	0.0638
0.25	0.5	0.0654
1.0	1.0	0.0691
2.25	1.5	0.0739
4	2	0.0795
6.25	2.5	0.0833
9.0	3.0	0.0868
12.25	3.5	0.0898
16	4	0.0922
20.25	4.5	0.0941
25	5	0.0954
36	6	0.0979
60	7.75	0.1004
120	10.95	0.1019
240	15.49	0.1029
480	21.91	0.1048
1440	37.95	0.1059

Table 16–1

CONSOLIDATION TEST

(Void ratio-pressure and coefficient of consolidation calculation)

Description of soil __Light brown clay__   Location __Cypress Circle__

Specimen diameter __2.5 in.__   Initial specimen height, $H_{t(i)}$ __1 in.__

Moisture content: Beginning of test __30.8__ (%) End of test __32.1__ %

Weight of dry soil specimen __116.74__ G_s __2.72__   Height of solids, H_s __1.356__   cm = __0.539 in.__

Pressure, p (ton/ft²)	Final dial reading (in.)	Change in specimen height (in.)	Final specimen height, $H_{t(f)}$ (in.)	Height of void, H_v (in.)	Final void ratio, e	Average height during consolidation, $H_{t(av)}$ (in.)	Fitting time (sec)		c_v from × 10³ (in.²/sec)	
(1)	(2)	(3)	(4)	(5)	(6)	(7)	t_{90} (8)	t_{50} (9)	t_{90} (10)	t_{50} (11)
0	0.0200		1.000	0.4610	0.855					
		0.0083				0.9959	302	68.7	0.696	0.711
1/2	0.0283		0.9917	0.4527	0.840					
		0.0073				0.9881	308	56.0	0.672	0.859
1	0.0356		0.9844	0.4454	0.826					
		0.0282				0.9703	492	144	0.406	0.322
2	0.0638		0.9562	0.4172	0.774					
		0.0421				0.9352	1102	294	0.131	0.147
4	0.1059		0.9141	0.3751	0.696					
		0.0455				0.8914	1354	240	0.124	0.163
8	0.1514		0.8686	0.3296	0.612					

Table 16-2

7. Determine the final specimen height, $H_{t(f)}$, at the end of consolidation due to a given loading (Col. 4 in Table 16–2). For example, in Table 16–2, $H_{t(f)}$ at $p = \frac{1}{2}$ ton/ft^2 is 0.9917. ΔH from $p = \frac{1}{2}$ ton/ft^2 and 1 ton/ft^2 is 0.0073 in. So, $H_{t(f)}$ at $p = 1$ ton/ft^2 is $0.9917 - 0.0073 = 0.9844$ in.

8. Determine the height of voids, H_v in the specimen at the end of consolidation due to a given loading, p, as (see Col. 5 in Table 16–2)

$$H_v = H_{t(f)} - H_s \tag{16.3}$$

9. Determine the final void ratio at the end of consolidation for each loading, p (see Col. 6, Table 16–2), as

$$e = \frac{H_v}{H_s} \tag{16.4}$$

10. Determine the average specimen height, $H_{t(av)}$, during consolidation for each incremental loading (Col. 7, Table 16–2). For example, in Table 16–2, the value of $H_{t(av)}$ between $p = \frac{1}{2}$ ton/ft^2 and $p = 1$ ton/ft^2 is

$$\frac{H_{t(f)} \text{ at } p = \frac{1}{2}\text{ton}/\text{ft}^2 + H_{t(f)} \text{ at } p = 1\text{ton}/\text{ft}^2}{2} = \frac{0.9917 + 0.9844}{2} = 0.9811 \text{ in.}$$

11. Calculate the coefficient of consolidation, c_v (Col. 10, Table 16–2), from t_{90} (Col. 8) as

$$T_v = \frac{c_v t}{H^2} \tag{16.5}$$

where T_v = time factor, $T_{90} = 0.848$
       H = maximum length of drainage path = $\dfrac{H_{t(av)}}{2}$
       (since the specimen is drained at top and bottom).

Thus

$$c_v = \frac{0.848 H_{t(av)}^2}{4 t_{90}} \tag{16.6}$$

$$\frac{tons}{ft^2} \times \frac{2000\,lb}{ton} = lb/ft^2$$

12. Calculate the coefficient of consolidation, c_v (Col. 11, Table 16–2), from t_{50} (Col. 9) as

$$T_{v\,(50\%)} = 0.197 = \frac{c_v t_{50}}{H^2} = \frac{c_v t_{50}}{\left[\dfrac{H_{t\,(av)}}{2}\right]^2}$$

$$c_v = \frac{0.197 H_{t\,(av)}^2}{4 t_{50}} \tag{16.7}$$

For example, from $p = \frac{1}{2}$ ton/ft^2 to $p = 1$ ton/ft^2,
$H_{t(av)} = 0.9881$ in.; t_{50} 56.0 sec;

$$c_v = \frac{0.197\,(0.9881)^2}{4\,(56)} = 0.859 \times 10^{-3}\ \text{in.}^2/\text{sec}$$

13. Plot a semilogarithmic graph of pressure vs. final void ratio (Col. 1 vs. Col. 6, Table 16–2). Pressure p is plotted on the log scale, and the final void ratio on the linear scale. As an example, the results of Table 16–2 are plotted in Fig. 16–6. *Note:* The plot has a curved upper portion and, after that, e vs. $\log p$ has a linear relationship.

14. Calculate the compression index, C_c. This is the slope of the linear portion of the e vs. $\log p$ plot (Step 13). In Fig. 16–6

$$C_c = \frac{e_1 - e_2}{\log \dfrac{p_2}{p_1}} = \frac{0.696 - 0.612}{\log \dfrac{8}{4}} = 0.279$$

15. On the semilogarithmic graph (Step 13), using the same horizontal scale (i.e., the scale for p), plot the values of c_v (Cols. 10 and 11, Table 16–2). As an example, the values determined in Table 16–2 are plotted in Fig. 16–6. *Note:* c_v is plotted on the linear scale corresponding to the average value of p, i.e.,

$$\frac{p_1 + p_2}{2}$$

16. Determine the *preconsolidation pressure*, p_c. The procedure can be explained with the aid of the e-$\log p$ graph drawn in Fig. 16–6 (Casagrande, 1936). First, determine point A, which is the point on the e-$\log p$ plot that has the smallest radius of curvature. Draw a horizontal line AB. Draw a line AD which is the *bisector* of the angle BAC. Project the straight line portion of the e-$\log p$ plot backwards to meet line AD at E. The pressure corresponding to point E is the preconsolidation pressure. In Fig. 16–6, $p_c = 1.6$ ton/ft^2.

General Comments

The magnitude of the compression index C_c varies from soil to soil. Many correlations for C_c have been proposed in the past for various types of soils. A summary of these correlations are given by Rendon-Herrero (1980). Following is a list of some of these correlations.

Correlation	Region of Applicability
$C_c = 0.007(LL - 7)$	Remolded clay
$C_c = 0.009(LL - 10)$	Undisturbed clays
$C_c = 1.15(e_0 - 0.27)$	All clays
$C_c = 0.0046(LL - 9)$	Brazilian clays
$C_c = 0.208e_0 + 0.0083$	Chicago clays

Note: LL = liquid limit;

e_0 = *in situ* void ratio

17
Triaxial Tests in Clay

Introduction

In Chapters 14 and 15, some aspects of the shear strength test for soil have been discussed. The triaxial compression test is a more sophisticated test procedure for determining the shear strength of soil. In general, with triaxial equipment, three types of common tests can be conducted. They are listed in Table 17–1. The unconsolidated-undrained test and the consolidated-undrained test mentioned in Table 17–1 will be described separately here.

Table 17–1. Common Triaxial Test in Clay

Type of Test	Parameters Determined
1. Unconsolidated-undrained (*U-U*)	c_u ($\phi = 0$)
2. Consolidated-drained (*C-D*)	c, ϕ
3. Consolidated-undrained (*C-U*)	$c, \phi, \overline{A}$

c_u = undrained cohesion
c = cohesion
ϕ = drained angle of friction
$\overline{A}$ = pore water pressure parameter.

Note: $s = c + \sigma' \tan \phi$ (c = cohesion, σ' = effective normal stress). For undrained condition, $\phi = 0$; $s = c_u$ [Eq. 15.2]

Equipment

1. Triaxial cell
2. Strain-controlled compression machine
3. Specimen trimmer
4. Wire saw

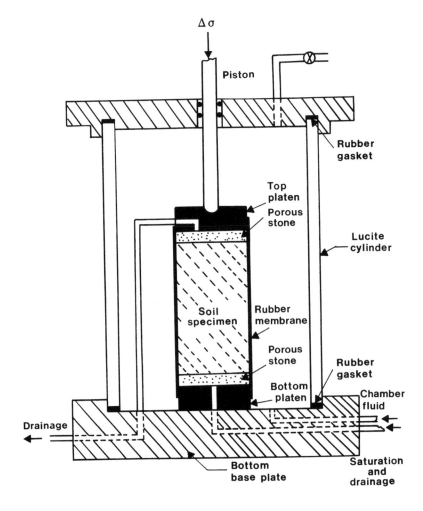

Figure 17-1. Schematic diagram of triaxial cell.

5. Vacuum source

6. Oven

7. Calipers

8. Evaporating dish

9. Rubber membrane

10. Membrane stretcher

Fig. 17–1 shows the schematic diagram of a triaxial cell. It mainly consists of a bottom base plate, a Lucite cylinder, and a top cover plate. A bottom platen is attached to the base plate. A porous stone is placed over the bottom platen, over which the soil specimen is placed. A porous stone and a platen are placed on top of the specimen. The specimen is enclosed inside a thin rubber membrane. Inlet and outlet tubes for specimen saturation and drainage are provided through the base plate. Appropriate valves to these tubes are attached to shut off the openings when desired. There is an opening in the base plate through which water (or glycerine) can be allowed to flow to fill the cylindrical chamber. A hydrostatic chamber pressure (σ_3) can be applied to the specimen through the chamber fluid. An added axial stress ($\Delta\sigma$) applied to the top of the specimen can be provided using a piston.

During the test, the triaxial cell is placed on the platform of a strain-controlled compression machine. The top of the piston of the triaxial chamber is attached to a proving ring. The proving ring is attached to a crossbar that is fixed to two metal posts. The platform of the compression machine can be raised (or lowered) at desired rates thereby raising (or lowering) the triaxial cell. During compression, the load on the specimen can be obtained from the proving ring readings and the corresponding specimen deformation from a dial gauge [1 small div = 0.001 in. (0.0254 mm)].

The connections to the soil specimen can be attached to a burette or a pore-water pressure measuring device to measure, respectively, the volume change of the specimen or the excess pore water pressure during the test.

Triaxial equipment is costly, depending on the accessories attached to it. For that reason, general procedures for tests will be outlined here. For detailed location of various components of the assembly, students will need the help of their instructor.

Triaxial Specimen

Triaxial specimens most commonly used are about 2.8 in. in diameter $\times$ 6.5 in. in height (71.1 mm dia $\times$ 165.1 mm height) or 1.4 in. in diameter $\times$ 3.5 in. in height (35.6 mm dia $\times$ 88.9 mm height). In any case, the length-to-diameter ratio should be between 2 and 3. For tests with undistributed natural soil samples collected in shelby tubes, a specimen trimmer may need to be used to prepare a specimen of desired dimensions. Depending on the triaxial cell at hand, for classroom use, remolded specimens can be prepared with Harvard miniature compaction equipment.

After the specimen is prepared, obtain its height (L_0) and diameter (D_0). The height should be taken 4 times about 90 degrees apart. The average of these four values should be equal to L_0. For the diameter, by using calipers, the measurement should be taken 4 times at the top, 4 times at the middle, and 4 times at the bottom of the specimen. The average values of these twelve measurements is D_0.

Placement of Specimen in the Triaxial Cell

1. Boil the two-porous stones to be used with the specimen.
2. De-air the lines connecting the base of the triaxial cell.
3. Attach the bottom platen to the base of the cell.
4. Place the bottom porous stone (moist) over the bottom platen.
5. Take a thin rubber membrane of appropriate size to fit the specimen tightly. Take a membrane stretcher, which is a brass tube with an inside diameter of about $\frac{1}{4}$ in. ($\approx$ 6 mm) larger than the specimen diameter (Fig. 17–2). The membrane stretcher can be connected to a vacuum source. Fit the membrane to the inside of the membrane stretcher, and lap the ends of the membrane over the stretcher. Then apply the vacuum. This will make the membrane form a smooth cover inside the stretcher.
6. Slip the soil specimen to the inside of the stretcher with the membrane (Step 5). The inside of the membrane may be moistened for ease of slipping the specimen in. Now release the vacuum, and unroll the membrane from the ends of the stretcher.

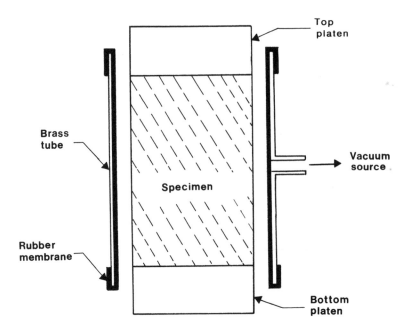

Figure 17-2. Membrane stretcher.

7. Place the specimen (Step 6) on the bottom porous stone (which is placed on the bottom platen of the triaxial cell), and stretch the bottom end of the membrane around the porous stone and bottom platen. At this time, place the top porous stone (moist) and the top platen on the specimen, and stretch the top of the membrane over it. For air-tight seals, it is always a good idea to apply some silicone grease around the top and bottom platens before the membrane is stretched over them.

8. Using some rubber bands, tightly fasten the membrane around the top and bottom platens.

9. Connect the drainage line leading from the top platen to the base of the triaxial cell.

10. Place the Lucite cylinder and the top of the triaxial cell on the base plate to complete the assembly.

   Note:

   1. In the triaxial cell, the specimen can be saturated by connecting the drainage line leading to the bottom of the specimen to a saturation reservoir. During this process, the drainage line leading from the top of the specimen is kept open to the atmosphere. The saturation of clay specimens takes a fairly long time.

   2. For unconsolidated-undrained test, if the specimen saturation is not required, nonporous plates can be used instead of porous stones at the top and bottom of the specimen.

Unconsolidated-Undrained Test

1. Place the triaxial cell (with the specimen inside it) on the platform of the compression machine.

2. Make proper adjustments so that the piston of the triaxial cell just rests on the top platen of the specimen.

3. Fill the chamber of the triaxial cell with water. Apply a hydrostatic pressure, σ_3, to the specimen through the chamber fluid. *Note:* All drainage to and from the specimen should be closed *now so that drainage from the specimen does not occur.*

4. Check for proper contact between the piston and the top platen on the specimen. Zero the dial gauge of the proving ring and the gauge used for measurement of the vertical compression of the specimen. Set the compression machine for a strain rate of about 0.5% per minute, and turn the switch on.

5. Take proving ring dial readings for vertical compression intervals of 0.01 in. (0.254 mm) initially. This interval can be increased to 0.02 in. (0.508 mm) or more later when the rate of increase of load on the specimen decreases. The proving ring readings will increase to a peak value and then may decrease or remain approximately constant. Take about four to five readings after the peak point.

6. After completion of the test, reverse the compression machine and lower the triaxial cell, and then shut off the machine. Release the chamber pressure, and drain the water in the triaxial cell. Then remove the specimen and determine its moisture content.

Calculation

The calculation procedure can be explained with reference to Tables 17–2a and b which present the results of an unconsolidated-undrained triaxial test on a dark brown silty clay specimen.

1. Calculate the final moisture content of the specimen, w, as (see Table 17–2a)

$$w\,(\%) \;=\; \frac{\text{moist weight of specimen} - \text{dry weight of specimen}}{\text{dry weight of specimen}}\,(100) \qquad (17.1)$$

2. Calculate the initial area of the specimen as (see Table 17–2a)

$$A_0 \;=\; \frac{\pi}{4}D_0^2 \qquad (17.2)$$

3. In Table 17–2b, Col. 2, calculate the vertical strain as

$$\epsilon \;=\; \frac{\Delta L}{L_0} \qquad (17.3)$$

where ΔL = total deformation of the specimen at any time.

UNCONSOLIDATED-UNDRAINED TRIAXIAL TEST
(Preliminary data)

Description of soil *Dark brown silty clay*

Location _____

Specimen No. *8*

Moist unit weight of specimen (end of test)	185.65 g
Dry unit weight of specimen	151.80 g
Moisture content (end of test)	22.3 %
Initial average length of specimen, L_0	3.52 in.
Initial average diameter of specimen, D_0	1.41 in.
Initial area, A_0	1.56 in.^{2}
G_s	2.73
Final degree of saturation	98.2 %
Cell confining pressure, σ_3	15 lb/in.^{2}
Proving ring calibration factor	0.37 lb/div.

Table 17-2a

UNCONSOLIDATED-UNDRAINED TRIAXIAL TEST
(Axial stress-strain calculation)

Specimen deformation $= \Delta L$ (in.) (1)	Vertical strain, $\epsilon = \dfrac{\Delta L}{L_0}$ (2)	Proving ring dial reading [No. of small divisions] (3)	Piston load, P [Col. 3 $\times$ calibration factor] (lb) (4)	Corrected area, $A = \dfrac{A_0}{1 - \epsilon}$ (in.^{2}) (5)	Deviatory stress, $\Delta \sigma = \dfrac{P}{A}$ (lb/in.^{2}) (6)
0	0	0	0	1.56	0
0.01	0.0028	3.5	1.295	1.564	0.828
0.02	0.0057	7.5	2.775	1.569	1.769
0.03	0.0085	11	4.07	1.573	2.587
0.04	0.0114	14	5.18	1.578	3.28
0.05	0.0142	18	6.66	1.582	4.210
0.06	0.0171	21	7.77	1.587	4.896
0.10	0.0284	31	11.47	1.606	7.142
0.14	0.0398	38	14.06	1.625	8.652
0.18	0.0511	44	16.28	1.644	9.903
0.22	0.0625	48	17.76	1.664	10.673
0.26	0.0739	52	19.24	1.684	11.425
0.30	0.0852	53	19.61	1.705	11.501
0.35	0.0994	52	19.24	1.735	11.109
0.40	0.1136	50	18.5	1.760	10.511
0.45	0.1278	49	18.13	1.789	10.134
0.50	0.1420	49	18.13	1.818	9.970

Table 17–2b

4. Calculate the piston load on the specimen (Col. 4, Table 17–2b) as

$$P = (\text{proving ring dial reading})(\text{calibration factor}) \qquad (17.4)$$

5. Calculate the corrected area, A, of the specimen (Table 17–2b, Col. 5) as

$$A = \frac{A_0}{1 - \epsilon} \qquad (17.5)$$

6. Calculate the deviatory stress (or piston stress), $\Delta\sigma$ (Table 17–2b, Col. 6), as

$$\Delta\sigma = \frac{P}{A} = \frac{\text{Col. 4}}{\text{Col. 5}} \qquad (17.6)$$

Graph

1. Draw a graph of the axial strain (%) vs. deviatory stress ($\Delta\sigma$). As an example, the results of Table 17–2b are plotted in Fig. 17–3. From this graph, obtain the value of $\Delta\sigma$ at failure ($\Delta\sigma = \Delta\sigma_f$).

2. The minor principal stress (*total*) on the specimen at failure is σ_3 (i.e., the chamber confining pressure). Calculate the major principal stress (*total*) at failure as

$$\sigma_1 = \sigma_3 + \Delta\sigma_f \qquad (17.7)$$

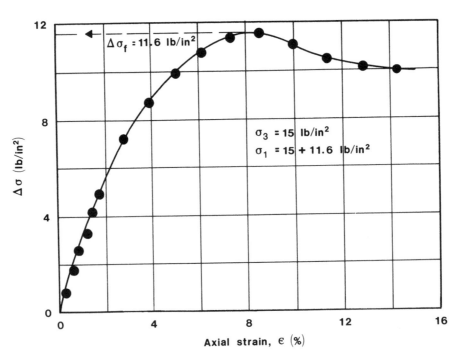

Figure 17–3. Plot of $\Delta\sigma$ against axial strain for the test reported in Table 17–2.

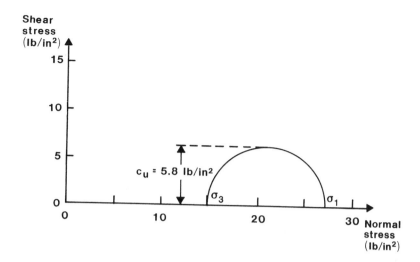

Figure 17–4. Total stress Mohr's circle at failure for test of Table 17–2 and Fig. 17–3.

3. Draw a Mohr's circle with σ_1 and σ_3 as the major and minor principal stresses. The radius of the Mohr's circle is equal to c_u. The results of the test reported in Table 17–2 and Fig. 17–3 are plotted in Fig. 17–4.

General Comments

1. For saturated clayey soils, the unconfined compression test (Chapter 15) is a special case of the U-U test discussed previously. For the unconfined compression test, $\sigma = 0$. However, the quality of results obtained from U-U tests is superior.

2. Fig. 17–5 shows the nature of the Mohr's envelope obtained from U-U tests with varying degrees of saturation. For *saturated specimens*, the value of $\Delta\sigma_f$, and thus c_u, is constant irrespective of the chamber confining pressure, σ. So, the Mohr's envelope is a horizontal line ($\phi = 0$). For soil specimens with degrees of saturation less than 100%, the Mohr's envelope is curved and falls above the $\phi = 0$ line.

Consolidated-Undrained Test

1. Place the triaxial cell with the saturated specimen on the compression machine platform and make adjustments so that the piston of the cell makes contact with the top platen of the specimen.

2. Fill the chamber of the triaxial cell with water, and apply the hydrostatic pressure, σ, to the specimen through the fluid.

3. The application of the chamber pressure, σ_3, will cause an increase of the pore water pressure in the specimen. For consolidation, connect the drainage lines from the specimen to a calibrated burette and leave the lines open. When the water level in the burette becomes constant, it will indicate that the consolidation is complete. For a saturated specimen, the volume change due to consolidation is equal to the volume of water drained into the burette. Record the volume of the drainage (ΔV).

4. Now connect the drainage lines to the pore-pressure measuring device.

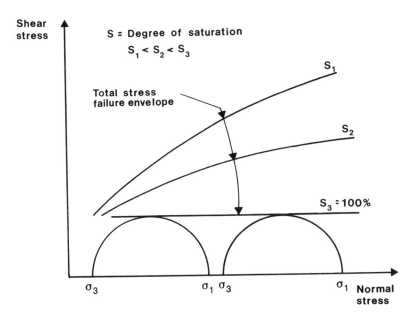

Figure 17-5. Nature of variation of total stress failure envelopes with the degree of saturation of soil specimen (for undrained test).

5. Check the contact between the piston and the top platen. Zero the proving ring dial gauge and the dial gauge, which measures the axial deformation of the specimen.

6. Set the compression machine for a strain rate of about 0.5% per minute, and turn the switch on. When the axial load on the specimen is increased, the pore water pressure in the specimen will also increase. Record the proving ring dial gauge reading and the corresponding excess pore water pressure (Δu) in the specimen for every 0.01 in. (0.254 mm) or less of axial deformation. The proving ring dial gauge reading will increase to a maximum and then decrease or remain approximately constant. Take at least four to five readings after the proving ring dial gauge reaches the maximum value.

7. At the completion of the test, reverse the compression machine and lower the triaxial cell. Shut off the machine. Release the chamber pressure (σ_3), and drain the water out of the triaxial cell.

8. Remove the tested specimen from the cell, and determine its moisture content.

9. Repeat the test on one or two more similar specimens. Each specimen should be tested at a different value of σ_3.

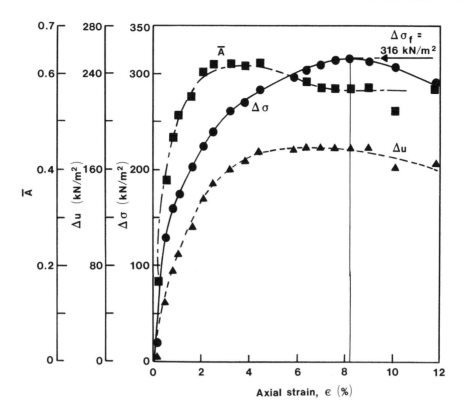

Figure 17-6. Plot of $\Delta\sigma$, Δu and $\overline{A}$ against axial strain for the consolidated drain test reported in Table 17–3.

Calculation and Graph

The procedure for making the required calculations and plotting graphs can be explained with reference to Table 17–3a and b and Figs. 17–6 and 17–7.

1. Determine the volume of the specimen after consolidation (Step 3) as

$$V_c = V_0 - \Delta V \tag{17.8}$$

where V_c = final volume of the specimen

$$V_0 = \text{initial volume} = \frac{\pi}{4} D_0^2 L_0 \tag{17.9}$$

2. Determine the length (L) and cross-sectional area of the specimen (A) after consolidation as

$$L_c = L_0 \left(\frac{V_c}{V_0}\right)^{1/3} \tag{17.10}$$

and

$$A_c = A_0 \left(\frac{V_c}{V_0}\right)^{2/3} \tag{17.11}$$

CONSOLIDATED-UNDRAINED TRIAXIAL TEST
(Preliminary data)

Description of soil   *Remolded grundite*

Location _____

Specimen No.   *2*

Moist unit weight of specimen (beginning of test)   _____

Moisture content (beginning of test)   *35.35%*

Initial length of specimen, L_0 *76.2 mm*

Initial diameter of specimen, D_0 *35.7 mm*

Initial area of the specimen, $A_0 = \frac{\pi}{4} D_0^2$ *10.0 cm²*

Initial volume of the specimen, $V_0 = A_0 L_0$ *76.2 cm³*

After Consolidation of Saturated Specimen in Triaxial Cell

Cell consolidation pressure, σ_3 *392 kN/m²*

Net drainage from the specimen during consolidation, ΔV *11.6 cm³*

Volume of specimen after consolidation, $V_0 - \Delta V = V_c$ *76.2 − 11.6 = 64.6 cm³*

Area of specimen after consolidation,

$$A_c = A_0 \left(\frac{V_c}{V_0} \right)^{2/3}$$

$10 \left(\dfrac{64.6}{76.2} \right)^{2/3} = 8.96 \ cm^3$

Length of the specimen after consolidation,

$$L_c = L_0 \left(\frac{V_c}{V_0} \right)^{1/3}$$

$76.2 \left(\dfrac{64.6}{76.2} \right)^{1/3} = 76.12 \ mm$
(2.84 in.)

Table 17–3a

CONSOLIDATED-UNDRAINED TRIAXIAL TEST
(Axial stress-strain calculation)

Proving ring calibration factor __1.0713 N/div__   (the results have been edited)

Specimen deformation = ΔL (in.) (1)	Vertical strain, $\epsilon = \dfrac{\Delta L}{L}$ (2)	Proving ring dial reading [No. of small divisions] (3)	Piston load, D [Col. 3 × calibration factor] (N) (4)	Corrected area = $A = \dfrac{A_c}{1-\epsilon}$ (cm²) (5)	Diviatory stress $\Delta\sigma = \dfrac{P}{A}$ (kN/m²) (6)	Excess pore water pressure, Δu (kN/m²) (7)	$\bar{A} = \dfrac{\Delta L}{\Delta\sigma}$ (8)
0	0	0	0	8.96	0	0	0
0.006	0.0021	15	16.07	8.98	17.90	2.94	0.164
0.015	0.0053	109	116.77	9.01	129.60	49.50	0.378
0.024	0.0085	135	144.63	9.04	159.99	74.56	0.466
0.030	0.0106	147	157.48	9.06	173.82	89.27	0.514
0.045	0.0158	172	184.26	9.11	202.26	111.83	0.553
0.060	0.0211	192	205.69	9.15	224.80	135.38	0.602
0.072	0.0254	205	219.62	9.19	238.98	148.13	0.620
0.090	0.0317	225	241.04	9.25	260.58	160.88	0.618
0.108	0.0380	236	252.83	9.31	271.57	167.75	0.618
0.124	0.0437	247	264.61	9.37	282.40	175.60	0.622
0.168	0.0592	265	283.89	9.52	298.20	176.58	0.592
0.180	0.0634	270	289.25	9.57	302.25	176.58	0.584
0.198	0.0697	278	297.82	9.63	309.26	176.58	0.570
0.216	0.0761	284	304.25	9.70	313.66	176.58	0.563
0.234	0.0824	287	307.46	9.76	315.02	176.58	0.561
0.257	0.0905	288	308.53	9.85	313.23	176.57	0.564
0.286	0.1007	286	306.39	9.96	307.62	160.88	0.523
0.335	0.1183	275	294.61	10.16	289.97	163.82	0.565

Table 17-3b

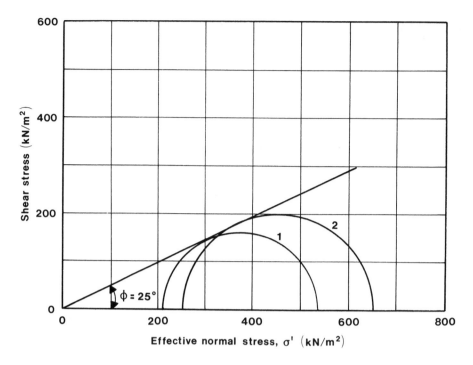

Figure 17-7. Effective stress Mohr's circles for remolded grundite reported in Table 17-3.

3. Calculate the axial strain (Table 17-3b, Col. 2) as

$$\epsilon = \frac{\Delta L}{L_c} \tag{17.12}$$

where ΔL = axial deformation.

4. Calculate the piston load, P (Table 17-3b, Col. 4)

$$P = (\text{proving ring dial reading}) (\text{calibration factor}) \tag{17.13}$$

5. Calculate the corrected area, A, as (see Table 17-3b, Col. 5)

$$A = \frac{A_c}{1 - \epsilon} \tag{17.14}$$

6. Determine the deviatory stress, $\Delta\sigma$ (Table 17-3b, Col. 6)

$$\Delta\sigma = \frac{P}{A} \tag{17.15}$$

7. Determine the pore water pressure parameter, $\bar{A}$

$$\bar{A} = \frac{\Delta u}{\Delta \sigma} \qquad (17.16)$$

where Δu = excess pore water pressure

8. Plot graphs for
   (a) $\Delta \sigma$ vs. ϵ (%)
   (b) Δu vs. ϵ (%)
   (c) $\bar{A}$ vs. ϵ (%)

   As an example, the results of the calculation shown in Table 17–3b have been plotted in Fig. 17–6.

9. From the $\Delta \sigma$ vs. ϵ (%) graph, determine the maximum value of $\Delta \sigma = \Delta \sigma_f$ and the corresponding values of $\Delta u = \Delta u_f$ and $A = A_f$.

   In Fig. 17–6, $\Delta \sigma_f = 316$ kN/m^2 at $\epsilon = 8.2\%$, and at the same strain level $\Delta u_f = 177$ kN/m^2 and $\bar{A} = 0.56$.

10. Calculate the *effective* major and minor principal stresses at failure.

    Effective minor principal stress at failure =

    $$\sigma_3 - \Delta u_f = \sigma'_3 \qquad (17.17)$$

    Effective major principal stress at failure =

    $$(\sigma_3 + \Delta \sigma_f) - \Delta u_f = \sigma'_1 \qquad (17.18)$$

    For the test on the remolded grundite reported in Table 17–3.

    $$\sigma'_3 = 392 - 177 = 215 \text{ kN/m}^2$$

    $$\sigma'_1 = (392 + 316) - 177 = 531 \text{ kN/m}^2$$

11. Collect σ'_1 and σ'_3 for all the specimens tested, and plot Mohr's circles. Plot a failure envelope that touches the Mohr's circles. The equation of the failure envelope can be given by

    $$s = c + \sigma' \tan \phi$$

    Determine the values of c and ϕ from the failure envelope.

    Fig. 17–7 shows the Mohr's circles for two tests on the remolded grundite reported in Table 17–3 (*Note:* The result for the Mohr's circle No. 2 is not given in Table 17–3). For the failure envelope, $c = 0$ and $\phi = 25°$. So

    $$s = \sigma' \tan 25°$$

General Comments

1. For normally consolidated soils $c = 0$; however, for over consolidated soils $c > 0$.

2. A typical range of values of $\overline{A}$ at failure for clayey soils is given below:

Type of Soil	$\overline{A}$ at Failure
Clays with high sensitivity	0.75 to 1.5
Normally consolidated clays	0.5 to 1.0
Overconsolidated clays	−0.5 to 0
Compacted sandy clay	0.5 to 0.75

References

1. American Society for Testing and Materials, *1991 Annual Book of ASTM Standards—Vol. 04.08*, Philadelphia, PA, 1991.

2. Atterberg, A., "Über die Physikalische Budenuntersuchung, and über die Plastizität der Töne," *Internationale Mitteilungen für Bodenkunde*, Vol. 1, 1911.

3. Casagrande, A., "Determination of Preconsolidation Load and Its Practical Significance," *Proceedings*, First International Conference on Soil Mechanics and Foundation Engineering, Vol. 3, 1936, pp. 60–64.

4. Casagrande, A. and Fadum, R. E., "Notes on Soil Testing for Engineering Purposes," *Engineering Publication No. 8*, Harvard University Graduate School, 1940.

5. Das, B. M., *Principles of Geotechnical Engineering, 2nd Edition*, PWS Engineering, Boston, 1990.

6. Proctor, R. R., "Design and Construction of Rolled Earth Dams," *Engineering News-Record*, August 31, September 7, September 21, and September 18, 1933.

7. Rendon-Herrero, O., "Universal Compression Index Equation," *Journal of the Geotechnical Engineering Division*, American Society of Civil Engineering, Vol. 106, No. GT11, 1980, pp. 1179–1200.

8. Taylor, D. W., "Research on the Consolidation of Clays," *Serial No. 82*, Department of Civil and Sanitary Engineering, Massachusetts Institute of Technology, 1942.

9. Waterways Experiment Station, "Simplification of the Liquid Limit Test Procedure," *Technical Memorandum No. 3–286*, 1949.

Appendix A
Engineering Classification of Soils

Introduction

In general, soils are widely varied in their grain-size distribution (Chapters 4 and 5). Also, depending on the type and quantity of clay minerals present, the plastic properties of soils (Chapters 5, 6, and 7) may be very different. For engineering works such as design of foundations and earth-retaining structures, construction of highways, and so on, it is necessary for different soils to be placed in specific groups and/or subgroups, based on their grain-size distribution and plasticity. Soil identification and classification in the field are necessary for various types of engineering works. The process of placing soils into various groups and/or subgroups is called *soil classification*.

For engineering purposes, there are two major systems presently used in the United States. They are: (i) the **American Association of State Highway and Transportation Officials (AASHTO) Classification System** and (ii) the **Unified Classification System**. These two systems will be elaborated on in the following sections.

A.1 American Association of State Highway and Transportation Officials (AASHTO) System of Classification

The AASHTO Classification System was originally initiated by the Highway Research Board (now called the Transportation Research Board) in 1943. It has gone through several changes since then. This system is presently used by federal, state, and county highway departments in the United States. In this soil classification system, soils are generally placed under seven major groups: **A-1, A-2, A-3, A-4, A-5, A-6**, and **A-7**. Group A-1 is divided into two subgroups, **A-1-a** and **A-1-b**. Group A-2 is divided into four subgroups: **A-2-4, A-2-5, A-2-6**, and **A-2-7**. Soils under Group A-7 are also divided into two subgroups. They are **A-7-5** and **A-7-6**. This system is also presently included in ASTM under Test Designation D-3284.

Along with the soil groups and subgroups discussed above, another factor called the **group index** (*GI*) is also included in this system. The importance of the group index can be explained in the following manner. Let us assume that two soils fall under the same group; however, they may have different values of *GI*. The soil that has a lower value of group index is likely to perform better as a highway subgrade material. The procedure for classifying soil under the AASHTO system is outlined below.

Step-by-Step Procedure for AASHTO Classification

1. Determine the percentage of soil passing through U.S. No. 200 sieve (0.075 mm opening).

   If 35% or less passes No. 200 sieve, it is a coarse-grained soil. Proceed to Steps 2 and 4.

   If more than 35% passes No. 200 sieve, it is a fine-grained material (i.e., silty or clayey material). For this, go to Steps 3 and 5.

Determination of Groups or Subgroups

2. For coarse-grained soils, determine the percent passing U.S. sieve Nos. 10, 40, and 200 and, also, the liquid limit and plasticity index. Then proceed to Table A–1. Start from the top line and compare the known soil properties with those given in the table (Columns 2 through 6). Go down one line at a time until a line is found for which all the properties of the desired soil matches. The soil group (or subgroup) is determined from Column 1.

3. For fine-grained soil, determine the liquid limit and the plasticity index. Then go to Table A–2. Start from the top line. By matching the soil properties from Columns 2, 3, and 4, determine the proper soil group (or subgroup).

Determination of Group Index

4. To determine the group index (*GI*) of coarse-grained soils, the following rules need to be followed.

   (a) *GI* for soils in groups (or subgroups) A-1-a, A-1-b, A-2-4, A-2-5, and A-3 is zero.

   (b) For *GI* of soils in groups A-2-6 and A-2-7, use the following equation:

   $$GI = 0.01\,(F_{200} - 15)\,(PI - 10) \tag{A.1}$$

   where F_{200} = percent passing No. 200 sieve and
   PI = plasticity index.

   If the *GI* comes out negative, round it off to zero. If the *GI* is positive, round it off to the nearest whole number.

5. For obtaining the *GI* of fine-grained soil, use the following equation:

   $$GI = (F_{200} - 35)\,[0.2 + 0.005\,(LL - 40)\,] + 0.01\,(F_{200} - 15)\,(PI - 10) \tag{A.2}$$

   If the *GI* comes out negative, round it off to zero. However, if it is positive, round it off to the nearest whole number.

Table A-1. AASHTO Classification for Coarse-Grained Soils

| Soil group (1) | | Grain size | | | Liquid limit* (5) | Plasticity index* (6) | Material type (7) | Subgrade rating (8) |
		Passing No. 10 sieve (2)	Passing No. 40 sieve (3)	Passing No. 200 sieve (4)				
A-1	A-1-a	50 max.	30 max.	15 max.		6 max.	Stone fragments, gravel, and sand	Excellent to good
	A-1-b		50 max.	25 max.		6 max.		
A-3			51 min.	10 max.		Nonplastic	Fine sand	
A-2	A-2-4			35 max.	40 max.	10 max.	Silty and clayey gravel and sand	
	A-2-5			35 max.	41 min.	10 max.		
	A-2-6			35 max.	40 max.	11 min.		
	A-2-7			35 max.	41 min.	11 min.		

* Based on the fraction passing No. 40 sieve

Table A–2. AASHTO Classification for Fine-Grained Soils

Soil group (1)		Passing No. 200 sieve (2)	Liquid limit* (3)	Plasticity index* (4)	Material type (5)	Subgrade rating (6)
A-4		36 min.	40 max.	10 max.	Silty soil	Fair to poor
A-5		36 min.	41 min.	10 max.	Silty soil	Fair to poor
A-6		36 min.	40 max.	11 min.	Clayey soil	Fair to poor
A-7	A-7-5	36 min.	41 min.	11 min. and $PI \leq LL - 30$	Clayey soil	Fair to poor
	A-7-6	36 min.	41 min.	11 min. and $PI > LL - 30$	Clayey soil	Fair to poor

* Based on the fraction passing U.S. No. 40 sieve

Expression for Soil Classification

6. The final classification of a soil is given by first writing down the group (or sub-group) followed by the group index in parenthesis.

General

Fig. A–1 shows the range of *PI* and *LL* for soil groups A-2-4, A-2-5, A-2-6, A-2-7, A-4, A-5, A-6, A-7-5, and A-7-6.

Example A-1

The following are the characteristics of two soils. Classify the soils according to the AASHTO system.

Soil A:	Percent passing No.	4 sieve	=	98
	Percent passing No.	10 sieve	=	90
	Percent passing No.	40 sieve	=	76
	Percent passing No.	200 sieve	=	34
	Liquid limit		=	38
	Plastic limit		=	26
Soil B:	Percent passing No.	4 sieve	=	100
	Percent passing No.	10 sieve	=	98
	Percent passing No.	40 sieve	=	86
	Percent passing No.	200 sieve	=	58
	Liquid limit		=	49
	Plastic limit		=	28

Solution

Soil A:

1. The soil has 34% (which is less than 35%) passing through No. 200 sieve. So this is a coarse-grained soil.

2. For this soil, liquid limit = 38.
   From Equation (7.2), plasticity index, $PI = LL - PL = 38 - 26 = 12$.
   From Table A–1, by matching, the soil is found to belong to subgroup **A-2-6**.

3. From Equation (A.1)
   $$GI = 0.01 \ (F_{200} - 15)(PI - 10)$$
   $$= 0.01 \ (34 - 15)(12 - 10) = (0.01)(19)(2)$$
   $$= 0.38 \approx 0$$

4. So, the soil can be classified as **A-2-6(0)**.

Soil B:

1. The soil has 58% (which is more than 35%) passing through No. 200 sieve. So this is a fine-grained soil.

2. The liquid limit of the soil is 49.
   From Equation (7.2), plasticity index, $PI = LL - PL = 49 - 28 = 21$.

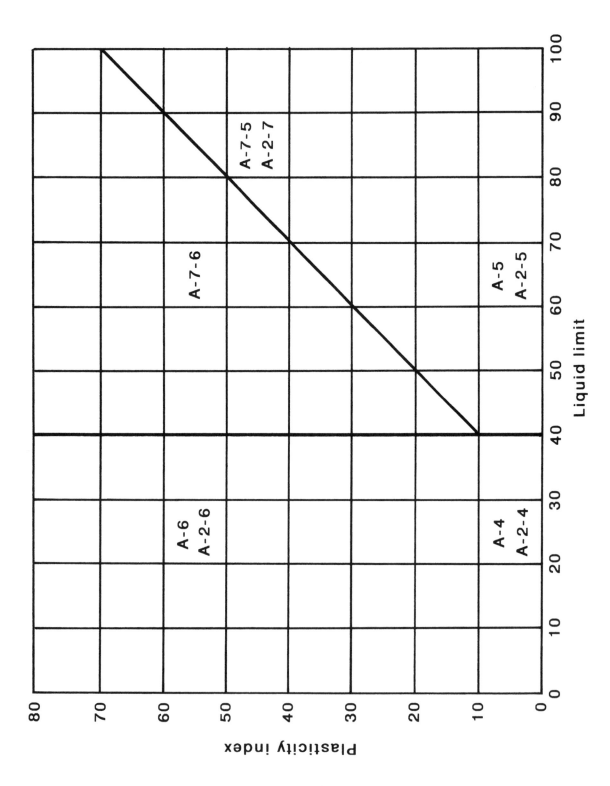

Figure A-1. Liquid limit and plasticity index for nine AASHTO soil groups.

3. From Table A–2, the soil is either A-7-5 or A-7-6.
   However, for this soil
   $$PI = 21 > LL - 30 = 49 - 30 = 19$$
   So the soil is **A-7-6**.

4. From Equation (A.2)
   $$\begin{aligned}GI \ &= (F_{200} - 35)\,[0.2 + 0.005\,(LL - 40)] + 0.01\,(F_{200} - 15)(PI - 10)\\ &= (58 - 35)\,[0.2 + 0.005\,(49 - 40)] + 0.01\,(58 - 15)(21 - 10)\\ &= 5.64 + 4.73 = 10.37 \approx 10\end{aligned}$$

5. So the soil is **A-7-6(10)**.

A.2 Unified Classification System

This classification system was originally started by Arthur Casagrande in 1942 for air-field construction during World War II. This work was conducted on behalf of the U.S. Army Corps of Engineers. At a later date, with the cooperation of the United States Bureau of Reclamation, the classification was modified. More recently, the American Society of Testing and Materials has introduced a more definitive system for the group name of soils. This system is also described in this section. In its present form, it is widely used by foundation engineers all over the world. Unlike the AASHTO system, the Unified system uses symbols to represent the soil types and the index properties of the soils. They are as follows:

Symbols to Represent Soil Type

Symbol	Soil Type
G	Gravel
S	Sand
M	Silt
C	Clay
O	Organic silts and clay
Pt	Highly organic soil and peat

Symbols to Represent Index Properties

Symbol	Index Property
W	Well-graded (for grain-size distribution)
P	Poorly-graded (for grain-size distribution)
L	Low to medium plasticity
H	High plasticity

Soil groups are developed by combining symbols for two categories listed above, such as **GW**, **SM**, and so forth.

Step-by-Step Procedure for Unified Classification System

1. If it is peat (i.e., primarily organic matter, dark in color, and has organic odor), classify it as **Pt** by visual observation. For all other soils, determine the percent of soil passing through U.S. No. 200 sieve (F_{200}).

2. Determine the percent retained on U.S. No. 200 sieve (R_{200}) as

$$R_{200} = 100 - F_{200} \qquad \text{(A.3)}$$

$\uparrow$

(nearest whole number)

3. If R_{200} is greater than 50%, it is a coarse-grained soil. However, if R_{200} is less than or equal to 50%, it is a fine-grained soil. If $R_{200} \leq 50\%$ (i.e., fine-grained soil), go to Step 4. However, if $R_{200} > 50\%$, go to Step 5.

4. For fine-grained soils (i.e., $R_{200} \leq 50\%$), determine if the soil is organic or inorganic in nature.

    (a) If the soil is organic, the group symbol can be **OH** or **OL**. If the soil is inorganic, the group symbol can be **CL, ML, CH, MH,** or **CL-ML**. (Fig. A–2 shows a plasticity chart with group symbols for fine-grained soils.)

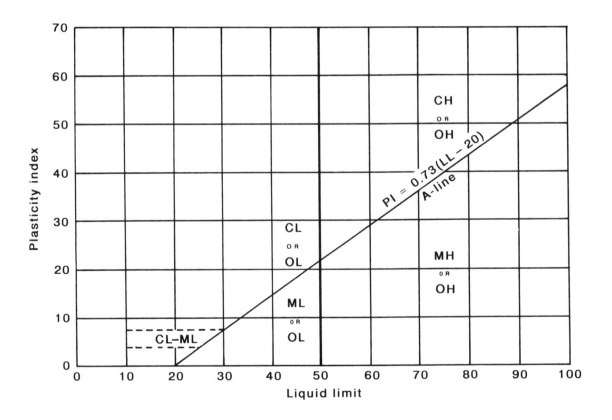

Figure A–2. Plasticity chart for group symbols of fine-grained soil.

(b) Determine the percent retained on U.S. No. 4 sieve, R_4, or

$$R_4 = 100 - F_4 \qquad \text{(A.4)}$$

$$\uparrow$$

(nearest whole number)

where F_4 = percent passing U.S. No. 4 sieve.

Note that R_4 is the percent of gravel fraction in the soil (GF), so

$$GF = R_4 \qquad \text{(A.5)}$$

(c) Determine the percent of sand fraction in the soil (SF), or

$$SF = R_{200} - GF \qquad \text{(A.6)}$$

(d) For inorganic soils, determine the liquid limit (LL) and the plasticity index (PI). Go to Step 4f.

(e) For organic soils, determine the liquid limit (not oven dried), LL_{NOD}; the liquid limit (oven dried), LL_{OD}; and the plasticity index (not oven dried), PI_{NOD}. Go to Step 4g.

(f) With known values of R_{200}, GF, SF, SF/GF, LL, and PI, use Table A–3 to obtain group symbols and group names of inorganic soils.

(g) With known value of LL_{NOD}, LL_{OD}, PI_{NOD}, R_{200}, GF, SF, and SF/GF, use Table A–4 to obtain group symbols and group names of organic soil.

5. For coarse-grained soils:

(a) If $R_4 > 0.5R_{200}$, it is a gravelly soil. These soils may have the following group symbols:

GW   GW-GM

GP   GW-GC

GM   GP-GM

GC   GP-GC

      GC-GM

Determine the following:

(1) F_{200}

(2) C_u = uniformity coefficient = D_{60}/D_{10} (see Chapter 4)

(3) C_c = coefficient of gradation = $D_{30}^2 / (D_{60} \times D_{10})$

(4) LL (of minus No. 40 sieve)

(5) PI (of minus No. 40 sieve)

(6) SF [based on Equations (A.3), (A.4), (A.5), and (A.6)]

Go to Table A–5 to obtain group symbols and group names.

Table A–3. Unified Classification of Fine-Grained Inorganic Soils
(*Note:* The group names are based on ASTM D-2487.)

Criteria for group symbol	Group symbol	Criteria for group name					Group name
		R_{200}	SF/GF	GF	SF		
$LL < 50$, $PI > 7$, and $PI > 0.73 (LL - 20)$	CL	<15	—	—	—		Lean clay
		15 to 29	≥1	—	—		Lean clay with sand
			<1	—	—		Lean clay with gravel
		≥30	≥1	<15	—		Sandy lean clay
			≥1	≥15	—		Sandy lean clay with gravel
			<1		<15		Gravelly lean clay
			<1		≥15		Gravelly lean clay with sand
$LL < 50$, $PI < 4$, or $PI < 0.73 (LL - 20)$	ML	<15	—	—	—		Silt
		15 to 29	≥1	—	—		Silt with sand
			<1	—	—		Silt with gravel
		≥30	≥1	<15	—		Sandy silt
			≥1	≥15	—		Sandy silt with gravel
			<1		<15		Gravelly silt
			<1		≥15		Gravelly silt with sand
$LL < 50$, $4 \leq PI \leq 7$, and $PI > 0.73 (LL - 20)$	CL-ML	<15	—	—	—		Silty clay
		15 to 29	≥1	—	—		Silty clay with sand
			<1	—	—		Silty clay with gravel
		≥30	≥1	<15	—		Sandy silty clay
			≥1	≥15	—		Sandy silty clay with gravel
			<1		<15		Gravelly silty clay
			<1		≥15		Gravelly silty clay with sand

continued

Table A–3. *continued*

Criteria for group symbol	Group symbol	Criteria for group name				Group name
		R_{200}	*SF/GF*	*GF*	*SF*	
$LL \geq 50$, and *PI* $\geq 0.73(LL - 20)$	CH	<15	—	—	—	Fat clay
		15 to 29	≥ 1	—	—	Fat clay with sand
			<1	—	—	Fat clay with gravel
		≥ 30	≥ 1	<15	—	Sandy fat clay
			≥ 1	≥ 15	—	Sandy fat clay with gravel
			<1		<15	Gravelly fat clay
			<1		≥ 15	Gravelly fat clay with sand
$LL \geq 50$, and *PI* $< 0.73 (LL - 20)$	MH	<15	—	—	—	Elastic silt
		15 to 29	≥ 1	—	—	Elastic silt with sand
			<1	—	—	Elastic silt with gravel
		≥ 30	≥ 1	<15	—	Sandy elastic silt
			≥ 1	≥ 15	—	Sandy elastic silt with gravel
			<1		<15	Gravelly elastic silt
			<1		≥ 15	Gravelly elastic silt with sand

Table A–4. Unified Classification of Fine-Grained Organic Soils
(*Note:* The group names are based on ASTM D-2487.)

Criteria for group symbol	Group symbol	Criteria for group name						
		Plasticity index	R_{200}	SF/GF	GF	SF	Group name
$LL_{NOD} < 50$, and $\dfrac{LL_{NOD}}{LL_{OD}} < 0.75$	OL	$PI_{NOD} \geqslant 4$ and $PI_{NOD} \geqslant 0.73 \times (LL_{NOD} - 20)$	<15	—	—	—	Organic clay
			15 to 29	$\geqslant 1$	—	—	Organic clay with sand
				<1	—	—	Organic clay with gravel
			$\geqslant 30$	$\geqslant 1$	<15	—	Sandy organic clay
				$\geqslant 1$	$\geqslant 15$	—	Sandy organic clay with gravel
				<1		<15	Gravelly organic clay
				<1		$\geqslant 15$	Gravelly organic clay with sand
		$PI < 4$ and $PI < 0.73 \times (LL_{NOD} - 20)$	<15	—	—	—	Organic silt
			15 to 29	$\geqslant 1$	—	—	Organic silt with sand
				<1	—	—	Organic silt with gravel
			$\geqslant 30$	$\geqslant 1$	<15	—	Sandy organic silt
				$\geqslant 1$	$\geqslant 15$	—	Sandy organic silt with gravel
				<1		<15	Gravelly organic silt
				<1		$\geqslant 15$	Gravelly organic silt with sand

continued

Table A–4. *continued*

Criteria for group symbol	Group symbol	Criteria for group name						
		Plasticity index	R_{200}	SF/GF	GF	SF	Group name
$LL_{NOD} \geq 50$, and $\dfrac{LL_{NOD}}{LL_{OD}} < 0.75$	OH	$PI_{NOD} \geq 0.73 \times (LL_{NOD} - 20)$	<15	—	—	—	Organic clay
			15 to 29	≥ 1	—	—	Organic clay with sand
				<1	—	—	Organic clay with gravel
			≥ 30	≥ 1	<15	—	Sandy organic clay
				≥ 1	≥ 15	—	Sandy organic clay with gravel
				<1		<15	Gravelly organic clay
				<1		≥ 15	Gravelly organic clay with sand
		$PI_{NOD} < 0.73 \times (LL_{NOD} - 20)$	<15	—	—	—	Organic silt
			15 to 29	≥ 1	—	—	Organic silt with sand
				<1	—	—	Organic silt with gravel
			≥ 30	≥ 1	<15	—	Sandy organic silt
				≥ 1	≥ 15	—	Sandy organic silt with gravel
				<1		<15	Gravelly organic silt
				<1		≥ 15	Gravelly organic silt with sand

Table A–5. Unified Classification of Gravelly Soils ($R_4 > 0.5 R_{200}$) (*Note*: The group names are based on ASTM D-2487.)

	Criteria for group symbol				Criteria for group name	
F_{200}	C_u	C_c	Relationship between LL & PI	Group symbol	SF	Group name
<5	≥ 4	$1 \leq C_c \leq 3$		GW	<15	Well graded gravel
					≥ 15	Well graded gravel with sand
	$C_u < 4$ and/or $1 > C_c > 3$			GP	<15	Poorly graded gravel
					≥ 15	Poorly graded gravel with sand
>12			$PI < 4$ or $PI < 0.73(LL - 20)$	GM	<15	Silty gravel
					≥ 15	Silty gravel with sand
			$PI > 7$ and $PI \geq 0.73(LL - 20)$	GC	<15	Clayey gravel
					≥ 15	Clayey gravel with sand
			$LL < 50$, $4 \leq PI \leq 7$ and $PI \geq 0.73(LL - 20)$	GC-GM	<15	Silty clayey gravel
					≥ 15	Silty clayey gravel with sand
$5 \leq F_{200} \leq 12$	≥ 4	$1 \leq C_c \leq 3$	$PI < 4$ or $PI < 0.73(LL - 20)$	GW-GM	<15	Well graded gravel with silt
					≥ 15	Well graded gravel with silt and sand
			$PI > 7$ and $PI > 0.73(LL - 20)$	GW-GC	<15	Well graded gravel with clay
					≥ 15	Well graded gravel with clay and sand
	$C_u < 4$ and/or $1 > C_c > 3$		$PI < 4$ or $PI < 0.73(LL - 20)$	GP-GM	<15	Poorly graded gravel with silt
					≥ 15	Poorly graded gravel with silt and sand
			$PI > 7$ and $PI > 0.73(LL - 20)$	GP-GC	<15	Poorly graded gravel with clay
					≥ 15	Poorly graded gravel with clay and sand

(b) If $R_4 \leq 0.5R_{200}$, it is a sandy soil. These soils may have the following group symbols:

SW	SW-SM
SP	SW-SC
SM	SP-SM
SC	SP-SC
SC-SM	

Determine the following:

(1) F_{200}

(2) $C_u = D_{60}/D_{10}$

(3) $C_c = D_{30}^2 / (D_{60} \times D_{10})$

(4) LL

(5) PI

(6) GF [based on Equations (A.3), (A.4), and (A.5)]

Go to Table A–6 to obtain group symbols and group names.

Example A-2

Classify Soils A and B as given in Example A-1 and obtain the group symbols and group names. Assume Soil B to be inorganic.

Soil A:
Percent passing No. 4 sieve = 98
Percent passing No. 10 sieve = 90
Percent passing No. 40 sieve = 76
Percent passing No. 200 sieve = 34
Liquid limit = 38
Plastic limit = 26

Soil B:
Percent passing No. 4 sieve = 100
Percent passing No. 10 sieve = 98
Percent passing No. 40 sieve = 86
Percent passing No. 200 sieve = 58
Liquid limit = 49
Plastic limit = 28

Solution

Soil A:

Step 1. $F_{200} = 34\%$

Step 2. $R_{200} = 100 - F_{200} = 100 - 34 = 66\%$

Step 3. $R_{200} = 66\% > 50\%$. So it is a coarse-grained soil.

Skip Step 4.

Step 5. $R_4 = 100 - F_4 = 2\%$
$R_4 < 0.5\,R_{200} = 33\%$

So it is a sandy soil (Step 5b). $F_{200} > 12\%$.

Table A-6. Unified Classification of Sandy Soils ($R_4 \le 0.5 R_{200}$)   (Note: The group names are based on ASTM D-2487.)

	Criteria for group symbol				Criteria for group name	
F_{200}	C_u	C_c	Relationship between LL & PI	Group symbol	GF	Group name
<5	≥ 6	$1 \le C_c \le 3$		SW	<15	Well graded sand
					≥ 15	Well graded sand with gravel
	$C_u < 6$ and/or $1 > C_c > 3$			SP	<15	Poorly graded sand
					≥ 15	Poorly graded sand with gravel
>12			$PI < 4$ or $PI < 0.73(LL - 20)$	SM	<15	Silty sand
					≥ 15	Silty sand with gravel
			$PI > 7$ and $PI \ge 0.73(LL - 20)$	SC	<15	Clayey sand
					≥ 15	Clayey sand with gravel
			$LL < 50$, $4 \le PI \le 7$ and $PI \ge 0.73(LL - 20)$	SM-SC	<15	Silty clayey sand
					≥ 15	Silty clayey sand with gravel
$5 \le F_{200} \le 12$	≥ 6	$1 \le C_c \le 3$	$PI < 4$ or $PI < 0.73(LL - 20)$	SW-SM	<15	Well graded sand with silt
					≥ 15	Well graded sand with silt and gravel
			$PI > 7$ and $PI > 0.73(LL - 20)$	SW-SC	<15	Well graded sand with clay
					≥ 15	Well graded sand with clay and gravel
	$C_u < 6$ and/or $1 > C_c > 3$		$PI < 4$ or $PI < 0.73(LL - 20)$	SP-SM	<15	Poorly graded sand with silt
					≥ 15	Poorly graded sand with silt and gravel
			$PI > 7$ and $PI > 0.73(LL - 20)$	SP-SC	<15	Poorly graded sand with clay
					≥ 15	Poorly graded sand with clay and gravel

So C_u and C_c values are not needed.

$$PI = LL - PL = 38 - 26 = 12$$
$$PI = 12 < 0.73 \, (LL - 20) = 0.73 \, (38 - 20) = 13.14$$

From Table A–6, the *group symbol* is **SM**.

$$GF = R_4 = 2\% \text{ (which is } < 15\%)$$

From Table A–6, the *group name is silty sand*.

Soil B:

Step 1. $F_{200} = 58\%$

Step 2. $R_{200} = 100 - F_{200} = 100 - 58 = 42\%$

Step 3. $R_{200} = 42\% < 50\%$. So it is a fine-grained soil.

Step 4. From Table A–3, $LL = 49 < 50$

$$PI = 49 - 28 = 21$$
$$PI = 21 < 73 \, (LL - 20) = 0.73 \, (49 - 20) = 21.17$$

So the *group symbol* is **ML**.

Again, $R_{200} = 42\% > 30\%$
$$R_4 = 100 - F_4 = 100 - 100 = 0\%$$

So $GF = 0\% < 15\%$
$$SF = R_{200} - GF = 42 - 0 = 42\%$$
$$SF/GF > 1$$

So the *group name is sandy silt*.

Appendix B
Computer Software

Two diskettes containing computer programs for analyzing data obtained from the laboratory tests described in this manual are attached to the back cover. These programs can be used to perform the necessary calculations and to prepare tables and the required graphs. The software package is interactive and offers menus that can be used to select any of the experiments. The programs generate tables in the same format as given in the manual. The user is prompted to enter data in specific columns of the tables on the screen. Calculated values are displayed along with the input test data on these tables. After data input is completed, the tables may be printed at the user's request. In the experiments that require graphic displays of results (e.g., the experiment on sieve analysis), the programs provide the necessary graphs on the screen and on the printer.

The computer programs are written in Advanced Basic for use on IBM and IBM-compatible personal computers. A graphics adapter and a printer capable of printing graphics are required. You may need to set your printer so that it can print a graphics screen by pressing SHIFT and PRINT SCREEN keys simultaneously. Refer to your printer manual for details.

Instructions for the Use of Computer Software:

1. Start DOS on your computer. To do so, insert your *DOS* diskette into drive A, turn on the computer, monitor, and printer, type the date and press ENTER, and type the time and press ENTER.
2. Type GRAFTABL, and press ENTER, to load graphics characters into the computer memory.
3. Type GRAPHICS and press ENTER. This enables the printer to print graphics display.
4. Type BASICA, and press ENTER, to start the Advanced Basic interpreter. If you are working on an IBM-compatible computer and have *GWBASIC*, remove the *DOS* diskette from drive A, insert the *GWBASIC* diskette, type GWBASIC, and press ENTER. For starting other versions of Advanced Basic, refer to your user's manual.

5. Remove the *DOS* (or Advanced Basic) diskette from drive A, insert the *Soil Mechanics Laboratory* diskette into drive A, type RUN"SOIL", and press ENTER. This will load the program "SOIL.BAS" into the computer memory and begin its execution.

6. The title will appear on the computer screen; press ENTER to continue. Next, the Software License Agreement will be displayed; press ENTER to continue.

7. The Main Menu, which lists all the experiments in the Manual, will then be presented. This main menu is shown in Fig. B-1. Use up (↑) and down (↓) arrow keys to move the cursor to the desired experiment (which will be highlighted), and press ENTER. (If you wish to end the computer session at this point, press the ESC key instead.)

8. The table for the selected experiment will be displayed on the screen, and the user will be prompted to enter data at the blinking cursor. Type in the requested data, check to make sure that it is correct; if not, then use the left (←) or right (→) arrow keys to move the cursor over the incorrect value, and type in the correct value. Then press ENTER. The program will perform the necessary calculations, and the calculated values will be displayed on the table along with the input data.

9. After you have entered all the data sets, terminate the data input mode by just pressing ENTER (or a blank) in place of the first input for the next data set.

10. The message "Do you want a PRINTOUT? (Y/N)" will then appear on the screen. Type Y for yes or N for no. (If you wish to end this experiment at this stage and return to the main menu to exit, press the ESC key instead.)

11. If the experiment requires graphs, the message "Do you have GRAPHICS Adapter? (Y/N)" will appear on the screen. If you do not have a graphics adapter, then type N or press the ESC key to return to the main menu. If you have a graphics adapter, then type Y, and the required graph will be displayed on the screen. To print the graph, press SHIFT (⇑) and PRTSC keys simultaneously. Fig. B-2 shows a typical plot of a graph plotted in this manner. It is an equivalent of Fig. 5–7 on page 32. You will need to use a French curve (or a straight edge, as the case may be) to complete the graph and obtain the desired quantity. [*Note:* The only exception to the procedure is Figure 6–5 (page 41), which is programmed to plot the flow line by using the least-square method. The liquid limit and the flow index are then programmed to be calculated.]

12. The main menu can be used to either run the computer program for another experiment (see Step 7) or end the computer session and return to DOS by pressing the ESC key.

13. Remove the *Soil Mechanics Laboratory* diskette from drive A, and turn off the computer.

Additional Comments

Since some of the computer programs in this software package require information from various figures in this manual, it may be necessary to have the manual at hand while using the software.

The computer software uses the same notation as is used in the manual, except that the capital delta (Δ) has been replaced by small delta (δ) in the computer programs.

```
* * * * * * * * * * * * * * * * * * * * * * * * * * * * * * * *
*                                                             *
*                      Computer Software                      *
*                            for                              *
*                                                             *
*           +-----------------------------------------+       *
*           | SOIL MECHANICS LABORATORY MANUAL        |       *
*           |                BY                       |       *
*           |            BRAJA DAS                    |       *
*           +-----------------------------------------+       *
*                                                             *
*                                                             *
*                      Prepared by:                           *
*                     Aslam Kassimali                         *
*         Department of Civil Engineering and Mechanics       *
*          Southern Illinois University at Carbondale         *
*                                                             *
* * * * * * * * * * * * * * * * * * * * * * * * * * * * * * * *
```

Press any key to continue.....

```
+-----------------------------------------------------------------+
|              Limited Use Software License Agreement             |
|              ------------------------------------              |
| Engineering Press, Inc. hereby grants the original purchaser a limited |
| license to use the software on a single computer, exclusively as an    |
| instructional aid for classroom use, in conjunction with the book      |
| SOIL MECHANICS LABORATORY MANUAL by Braja Das. Any other use is        |
| illegal.                                                               |
| You may make upto three backup or archival copies of the original      |
| diskette for the purpose of running the software. You agree that no    |
| other copies will be made. You acknowledge that the software is the    |
| sole and exclusive property of Engineering Press, Inc. You do not      |
| become the owner of the software, but do have the right to use it in   |
| accordance with the agreement. You may not transfer your license to    |
| use the software without the prior written consent of Engineering      |
| Press, Inc.                                                            |
| Engineering Press, Inc., Braja Das, and Aslam Kassimali will not be    |
| responsible for any direct, incidental, or consequential damages       |
| resulting from the use of this software. If any of the terms and       |
| conditions of this agreement are broken, Engineering Press, Inc. has   |
| the right to terminate the agreement and demand that you return the    |
| software.                                                              |
+-----------------------------------------------------------------+
```

Press any key to continue.....

```
┌─────────────────────────────────────────────────────────────────────────┐
│                                                                           │
│                        ┌───────────────────────┐                          │
│                        │  M A I N - M E N U    │                          │
│                        └───────────────────────┘                          │
│                                                                           │
│           Determination of Moisture Content                               │
│           Specific Gravity                                                │
│           Sieve Analysis                                                  │
│           Hydrometer Analysis                                             │
│           Liquid Limit Test                                               │
│           Plastic Limit Test                                              │
│           Shrinkage Limit Test                                            │
│           Constant Head Permeability Test in Sand                         │
│           Falling Head Permeability Test in Sand                          │
│           Standard Proctor Compaction Test                                │
│           Modified Proctor Compaction Test                                │
│           Determination of Field Unit Weight of Compaction by Sand Cone Method │
│           Direct Shear Test on Sand                                       │
│      ══►  Unconfined Compression Test                                     │
│           Consolidation Test                                              │
│           Triaxial Tests in Clay                                          │
│                                                                           │
│                                                                           │
│ Use ↑ or ↓ to select a topic, then press ◄─┘        Esc=Exit to DOS      │
│                                                                           │
└─────────────────────────────────────────────────────────────────────────┘
```

Figure B-1. The main menu.

```
┌─────────────────────────────────────────────────────────────────┐
│ D E T E R M I N A T I O N - O F - M O I S T U R E - C O N T E N T │
└─────────────────────────────────────────────────────────────────┘
```

Description of soil: Silty clay
Location: Southern Illinois University at Carbondale
Sample No.: 34

No. of can	Weight of can, W1 (g)	Weight of can & wet soil, W2 (g)	Weight of can & dry soil, W3 (g)	$w(\%)=\dfrac{W2-W3}{W3-W1}\times100$
46	16.09	31.32	29.28	15.47
47	100.50	125.03	119.00	32.59
48	5.80	7.25	6.93	28.32
105	16.09	31.32	29.28	15.47

```
┌───────────────────────────────────┐
│ S P E C I F I C - G R A V I T Y │
└───────────────────────────────────┘
```

Description of soil: Light brown sandy silt
Location: Southern Illinois University at Carbondale
Sample No.: 134
Volume of flask at 20°C (ml): 500

Test No.:	1	2
Volumetric flask no.:	6	67
Weight of flask + water filled to mark, W1 (g):	660.00	660.00
Weight of flask + soil + water filled to mark, W2 (g):	722.00	722.00
Weight of dry soil, W3 (g):	99.00	99.00
Temperature of test, T1 °C:	23.00	20.00
$Gs\ (at\ T1\ °C) = \dfrac{W3}{(W1+W3)-W2}$	2.676	2.676
A (from Table 3-2):	0.9993	1.0000
Gs (at 20°C) = Gs (at T1°C) x A	2.674	2.676

S I E V E - A N A L Y S I S

Description of soil: Sand with some fines
Location: Southern Illinois University at Carbondale
Sample No.: 134   Weight of oven dry sample, W (g):   500

Sieve No.	Sieve opening (mm)	Weight retained on each sieve (g)	Percent of weight retained on each sieve	Cumulative percent retained	Percent finer
4	4.750	0.00	0.00	0.00	100.00
10	2.000	40.20	8.04	8.04	91.96
20	0.850	84.60	16.92	24.96	75.04
40	0.425	90.20	18.04	43.00	57.00
60	0.250	106.40	21.28	64.28	35.72
140	0.106	108.80	21.76	86.04	13.96
200	0.075	59.40	11.88	97.92	2.08
Pan	---	8.70			

Σ 498.30 = W1

Loss during sieve analysis=[(W-W1)/W]x100= 0.34% (OK if less than 2%)

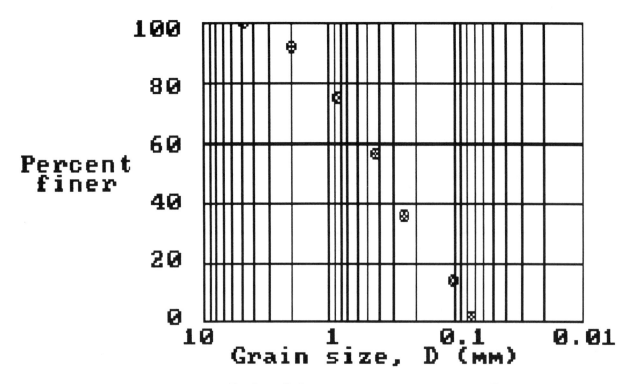

Plot of percent finer vs. grain size

Figure B-2. Sample graph printout.

H Y D R O M E T E R - A N A L Y S I S

Description of soil: Brown Silty Clay
Sample No.: 134   Location: Carbondale, Illinois
Gs: 2.75   Hydrometer type: ASTM 152H
Dry weight of soil, Ws (g): 50 Temperature of test, T (°C): 28
Meniscus correction, Fm: 1   Zero correction, Fz: 7
Temperature correction, FT: 2.15

Time (min)	Hydrometer reading, R	Rcp	Percent finer, aRcp ----x100 50	Rcl	L (Fig 5-2) (cm)	A (Fig 5-3)	D (mm)
0.25	51	46.15	90.31	52	7.80	0.0121	0.0676
0.50	48	43.15	84.44	49	8.30	0.0121	0.0493
1.00	47	42.15	82.48	48	8.40	0.0121	0.0351
2.00	46	41.15	80.53	47	8.60	0.0121	0.0251
4.00	45	40.15	78.57	46	8.80	0.0121	0.0179
8.00	44	39.15	76.61	45	8.95	0.0121	0.0128
15.00	43	38.15	74.65	44	9.10	0.0121	0.0094
30.00	42	37.15	72.70	43	9.25	0.0121	0.0067
60.00	40	35.15	68.78	41	9.60	0.0121	0.0048
120.00	38	33.15	64.87	39	9.90	0.0121	0.0035
240.00	34	29.15	57.04	35	10.50	0.0121	0.0025
480.00	32	27.15	53.13	33	10.90	0.0121	0.0018
1440.00	29	24.15	47.26	30	11.35	0.0121	0.0011
2880.00	27	22.15	43.34	28	11.65	0.0121	0.0008

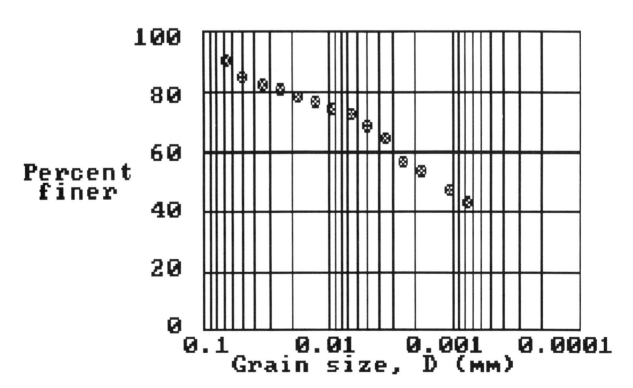

Plot of percent finer vs. grain size

L I Q U I D - L I M I T - T E S T

Description of soil: Gray silty clay
Location: Southern Illinois University, Carbondale, Illinois
Sample No.:   1250

Can No.	Weight of can, W1 (g)	Weight of can & wet soil, W2 (g)	Weight of can & dry soil, W3 (g)	Moisture content, w (%)	Number of blows, (N)
8	15.26	29.30	25.84	32.70	35
21	17.01	31.58	27.72	36.04	23
25	15.17	31.45	26.96	38.08	17

Liquid limit = 35.28
Flow index = 17.23

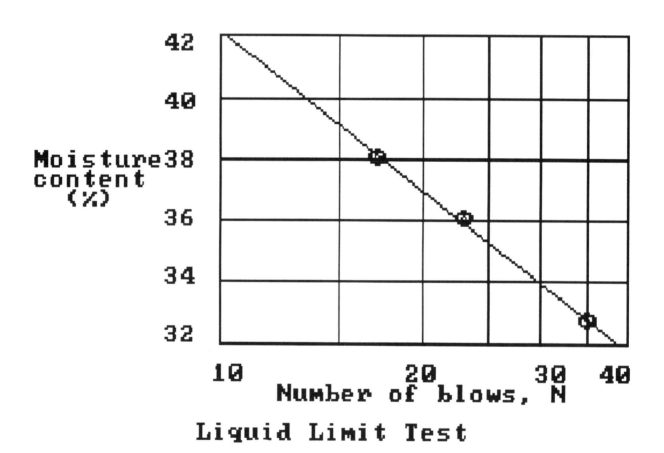

Liquid Limit Test

PLASTIC - LIMIT - TEST

Description of soil: Gray clayey silt
Location: Southern Illinois University at Carbondale
Sample No.: 34

Can No.	Weight of can, W1 (g)	Weight of can & wet soil, W2 (g)	Weight of can & dry soil, W3 (g)	$PL = \dfrac{W2-W3}{W3-W1} \times 100$
103	13.33	23.86	22.27	17.79
46	16.09	31.32	29.28	15.47
47	100.50	125.03	119.00	32.59
48	5.80	7.25	6.93	28.32

SHRINKAGE - LIMIT - TEST

Description of soil: Dark brown clay
Location: Lake Drive
Sample No.: 129

Test No.	1	13
Weight of coated shrinkage limit dish, W1 (g)	12.34	12.34
Weight of dish + wet soil, W2 (g)	40.43	40.43
Weight of dish + dry soil, W3 (g)	33.68	33.68
$wi\ (\%) = [(W2-W3)/(W3-W1)] \times 100$	31.63	31.63
Weight of mercury to fill the dish, W4 (g)	198.83	198.83
Weight of mercury displaced by soil pat, W5 (g)	150.30	150.30
$\dfrac{(W4-W5)100}{(13.6)(W3-W1)}$	16.72	16.72
$SL = wi - \dfrac{(W4-W5)100}{(13.6)(W3-W1)}$	14.91	14.91

CONSTANT - HEAD - PERMEABILITY - TEST

Description of soil: Uniform Sand
Sample No.: 14   Location: Southern Illinois
Length of specimen,L (cm):13.20 Diameter of specimen,D (cm): 6.35
Area of specimen,A (cm²):31.67 Dry weight of specimen,W2-W1 (g):726.90
Void ratio,e: 0.53   Specific gravity,Gs: 2.66

Test No.	Average flow, Q (cu. cm)	Time of collection, t (sec)	Temperature of water, T (°C)	Head difference, h (cm)	$k = \dfrac{QL}{Aht}$ (cm/sec)
1	305.00	60.00	25.00	60.00	0.035
2	375.00	60.00	25.00	70.00	0.037
3	395.00	60.00	25.00	80.00	0.034

k(20°C) = 0.0317 cm/sec   Average k = 0.036 cm/sec

FALLING - HEAD - PERMEABILITY - TEST

Description of soil: Uniform Sand
Sample No.: 52   Location: Carbondale
Length of specimen,L (cm):13.20   Area of specimen,A (cm²):31.65
Dry weight of specimen (g):726.90   Gs: 2.66 Void ratio,e: 0.53

Test No.	h1 (cm)	h2 (cm)	Test duration, t (sec)	Temperature of water, T (°C)	Volume of water, V (cu. cm)	k Eq. (10.3) (cm/sec)
1	85.00	24.00	15.40	25.00	64.00	0.036
2	75.00	20.00	15.30	25.00	58.00	0.038
3	65.00	20.00	14.40	25.00	47.00	0.036

k(20°C) = 0.0325 cm/sec   Average k = 0.037 cm/sec

P R O C T O R - C O M P A C T I O N - T E S T

Description of soil: Light brown clayey silt
Sample No.: 134   Location: Carbondale, Illinois
Volume of mold: 1/30   Weight of hammer (lb): 5
No. of blows/layers: 25 No. of layers: 3   Gs: 2.68

Test No.	Weight of mold, W1 (lb)	Weight of mold + moist soil,W2 (lb)	Weight of moist soil, W2 - W1 (lb)	Moist unit weight, 30(W2-W1) (lb/cu ft)	Moisture content, w (%)	Dry unit weight, rd (lb/cu ft)
1	10.348	14.185	3.837	115.110	8.743	105.855
2	10.348	14.407	4.059	121.770	10.269	110.429
3	10.348	14.526	4.178	125.340	10.924	112.996
4	10.348	14.629	4.281	128.430	12.516	114.144
5	10.348	14.509	4.161	124.830	15.036	108.514
6	10.348	14.466	4.118	123.540	18.732	104.050

MOISTURE CONTENT DETERMINATION

Test No.:	1	2	3	4	5	6
Can No.:	202	212	222	242	206	504
Weight of can (g):	54.00	53.25	53.25	54.00	54.75	40.80
Weight of can + moist soil (g):	253.00	354.01	439.00	490.00	422.83	243.00
Weight of can + dry soil (g):	237.00	326.00	401.01	441.50	374.72	211.10
Moisture content (%):	8.74	10.27	10.92	12.52	15.04	18.73

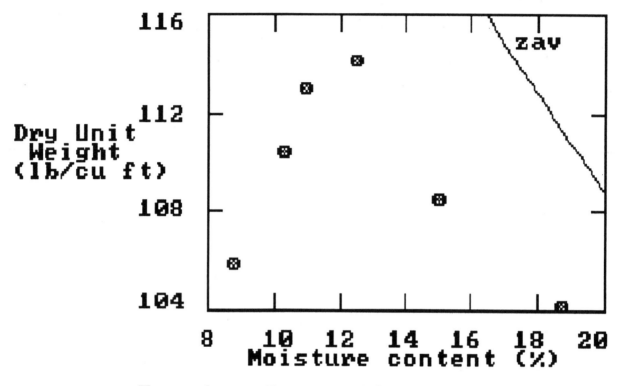

Proctor Compaction Test

FIELD UNIT WEIGHT - SAND CONE METHOD

CALIBRATION OF UNIT WEIGHT OF OTTAWA SAND

    Weight of sand in the mold, W1 (lb):   3.31

    Volume of mold, V1 (cu ft):   0.03

    rd(sand) = W1/V1 (lb/cu ft):   99.30

CALIBRATION CONE

    Weight of bottle + cone + sand (before use), W2 (lb): 15.17

    Weight of bottle + cone + sand (after use), W3 (lb): 14.09

    Weight of sand to fill the cone, Wc=W2-W3 (lb):   1.08

RESULTS FROM FIELD TESTS

    Weight of bottle + cone + sand (before use), W5 (lb): 15.42

    Weight of bottle + cone + sand (after use), W7 (lb): 11.74

    Volume of hole, [W5-W7-Wc]/rd(sand) (cu ft):   0.0262

    Weight of gallon can, W4 (lb):   0.82

    Weight of can + moist soil, W6 (lb):   3.92

    Weight of can + dry soil, W8 (lb):   3.65

    Weight of moist soil, W6-W4 (lb):   3.10

Moist unit weight of soil in field,
  r = [W6-W4]rd(sand)/[W5-W7-Wc] (lb/cu ft):   118.40

Moisture content in the field,
  w(%) = (W6-W8)/(W8-W4) x 100 :   9.54

$$\text{Dry unit weight in the field}, rd = \frac{r}{1+(w\%/100)} \text{(lb/cu ft)}: 108.08$$

DIRECT - SHEAR - TEST - ON - SAND

Description of soil: Uniform sand
Specimen No.: 1985   Location:Carbondale, Illinois
Size of specimen-Length,L(in):2.00 Width,B(in):2.00 Height,H(in):1.31
Weight of specimen, W(g):143.10   Dry unit weight (lb/cu ft):104.03
Gs: 2.66   Void ratio, e: 0.595   Normal load, N(lb): 56.00
Normal stress, σ'(lb/in²): 14.00
Proving ring calibration factor (lb/div.): 0.310

Horizontal displacement (in)	Vertical displacement* (in)	No. of div. in proving ring dial gauge	Shear force, S (lb)	Shear stress, τ (lb/in²)
0.000	0.0000	0	0.000	0.000
0.010	0.0010	45	13.950	3.488
0.020	0.0020	76	23.560	5.890
0.030	0.0040	95	29.450	7.363
0.040	0.0060	112	34.720	8.680
0.050	0.0080	124	38.440	9.610
0.060	0.0090	129	39.990	9.998
0.070	0.0100	125	38.750	9.688
0.080	0.0100	119	36.890	9.223
0.090	0.0090	114	35.340	8.835
0.100	0.0080	109	33.790	8.448
0.110	0.0080	108	33.480	8.370
0.120	0.0080	105	32.550	8.138

* + sign means expansion

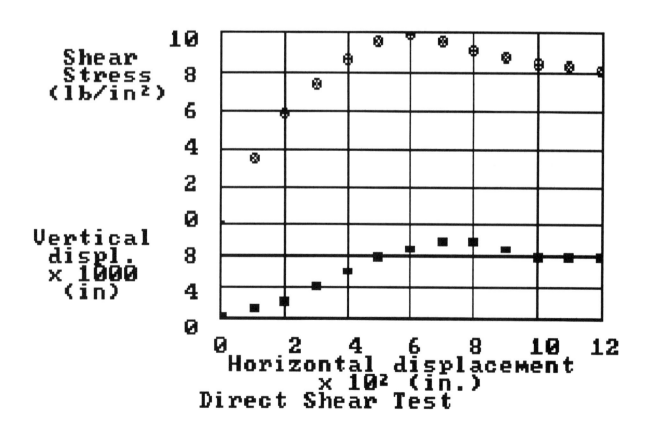

Direct Shear Test

```
┌─────────────────────────────────────────────────────┐
│ U N C O N F I N E D - C O M P R E S S I O N - T E S T │
└─────────────────────────────────────────────────────┘
```

Description of soil: Light brown clay
Specimen No.: 124   Location: Carbondale, Illinois
Moist weight of specimen(g):149.8 Moisture content(%): 12.0
Length of specimen,L(in): 3 Diameter,D(in):1.35   Area,Ao(in²): 1.43
Proving ring calibration factor: 1 div. = 0.264 lb

Specimen deformation = δL (in)	Vertical strain, $\epsilon = \dfrac{\delta L}{L}$	Proving ring dial reading -No. of small divisions	Load=Col. 3x calibration factor of proving ring (lb)	Corrected area=Ac=$\dfrac{Ao}{1-\epsilon}$ (in²)	Stress = $\dfrac{Col.\ 4}{Col.\ 5}$ (lb/in²)
0.00	0.0000	0	0.000	1.431	0.000
0.01	0.0033	12	3.168	1.436	2.206
0.02	0.0067	38	10.032	1.441	6.962
0.03	0.0100	52	13.728	1.446	9.495
0.04	0.0133	58	15.312	1.451	10.555
0.06	0.0200	67	17.688	1.461	12.110
0.08	0.0267	74	19.536	1.471	13.284
0.10	0.0333	78	20.592	1.481	13.907
0.12	0.0400	81	21.384	1.491	14.342
0.14	0.0467	83	21.912	1.501	14.594
0.16	0.0533	85	22.440	1.512	14.841
0.18	0.0600	86	22.704	1.523	14.910
0.20	0.0667	86	22.704	1.534	14.804
0.24	0.0800	84	22.176	1.556	14.253
0.28	0.0933	83	21.912	1.579	13.879
0.32	0.1067	82	21.648	1.602	13.511
0.36	0.1200	81	21.384	1.627	13.147

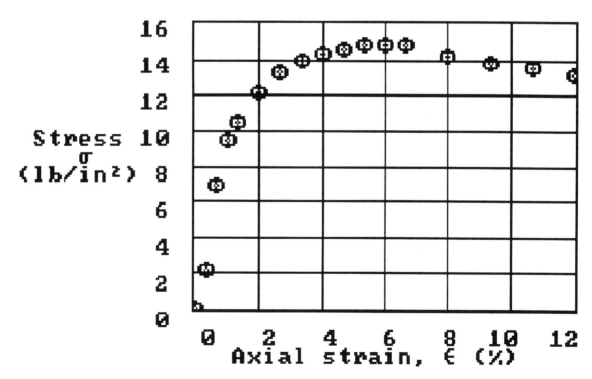

Unconfined Compression Test

CONSOLIDATION-TEST

(Time vs. vertical dial reading)

Description of soil: Light brown clay
Pressure on specimen (ton/ft²): 4.000
Clock time of load application: 8:35 am

Time after load application, t (min)	√t (√min)	Vertical dial reading (in)
0.00	0.00	0.063800
0.25	0.50	0.065400
1.00	1.00	0.069100
2.25	1.50	0.073900
4.00	2.00	0.079500
6.25	2.50	0.083300
9.00	3.00	0.086800
12.25	3.50	0.089800
16.00	4.00	0.092200
20.25	4.50	0.094100
25.00	5.00	0.095400
36.00	6.00	0.097900
60.00	7.75	0.100400
120.00	10.95	0.101900
240.00	15.49	0.102900
480.00	21.91	0.104800
1440.00	37.95	0.105900

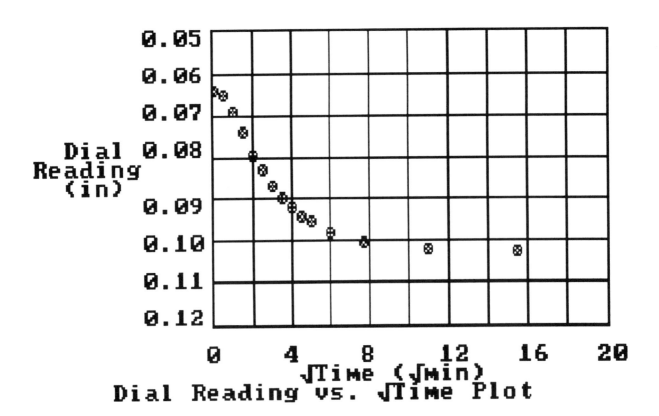

Dial Reading vs. √Time Plot

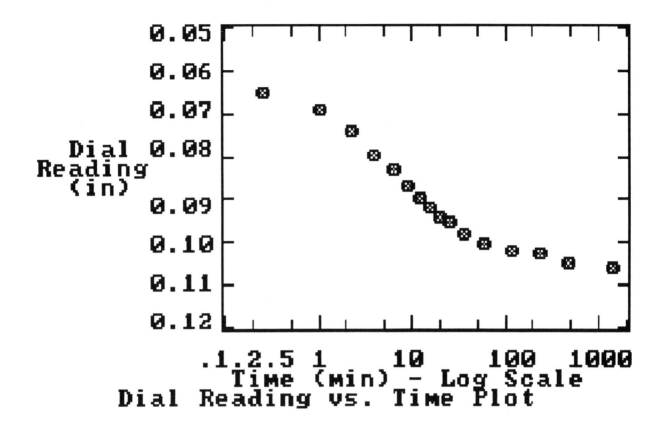

$$\boxed{\text{C O N S O L I D A T I O N - T E S T}}$$

(Void ratio-pressure and coefficient of consolidation calculation)

Description of soil:Light brown clay   Location:Southern Illinois
Specimen diameter (in): 2.50   Initial specimen height,Ht(i) (in):1.00
Moisture content: beginning of test(%): 30.8   End of test(%): 32.1
Weight of dry soil specimen,(g):116.74Gs:2.72 Height of solids,Hs(in):0.53

Pres-sure, P	Final dial read-ing	Change in speci-men height	Final speci-men height Ht(f)	Height of void Hv	Final void ratio, e	Average height during consol-idation Ht(av)	Fitting time (sec)		Cv x 1000 (in²/sec)	
T/ft²	(in)	(in)	(in)	(in)		(in)	t90	t50	t90	t50
0.00	0.0200		1.0000	0.4662	0.8734					
		0.0083				0.9958	302	69	0.696	0.711
0.50	0.0283		0.9917	0.4579	0.8579					
		0.0073				0.9881	308	56	0.672	0.859
1.00	0.0356		0.9844	0.4506	0.8442					
		0.0282				0.9703	492	144	0.406	0.322
2.00	0.0638		0.9562	0.4224	0.7914					
		0.0421				0.9352	1102	294	0.168	0.146
4.00	0.1059		0.9141	0.3803	0.7125					
		0.0455				0.8914	1354	240	0.124	0.163
8.00	0.1514		0.8686	0.3348	0.6272					

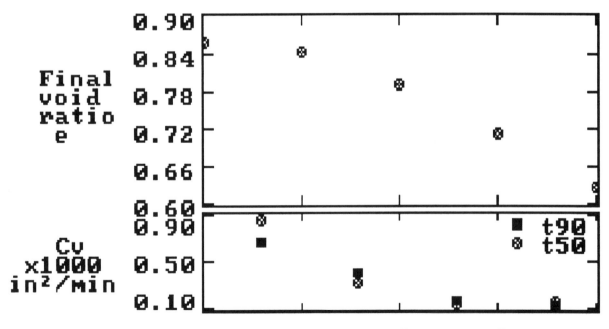

Plot of void ratio and the coefficient of consolidation against pressure

UNCONSOLIDATED-UNDRAINED TRIAXIAL TEST

(Preliminary data)

Description of soil: Dark brown silty clay
Location: Carbondale, Illinois
Specimen No.: 139

Moist unit weight of specimen (end of test) (g):	185.65
Dry unit weight of specimen (g):	151.80
Moisture content (end of test) (%):	22.30
Initial average length of specimen, Lo (in):	3.52
Initial average diameter, Do (in):	1.41
Initial area, Ao (in^2):	1.56
Gs :	2.73
Final degree of saturation (%):	98.20
Cell confining pressure, $\sigma 3$ (lb/in^2):	15.00
Proving ring calibration factor (lb/div):	0.37

(Axial stress-strain calculation)

Specimen deformation = δL (in)	Vertical strain, $\epsilon = \dfrac{\delta L}{Lo}$	Proving ring dial reading -No. of small divisions	Piston load, P = Col. 3 x calibration factor (lb)	Corrected area=A=$\dfrac{Ao}{1-\epsilon}$ (in^2)	Deviatory stress $\delta\sigma = P/A$ (lb/in^2)
0.00	0.0000	0.00	0.000	1.561	0.000
0.01	0.0028	3.50	1.295	1.566	0.827
0.02	0.0057	7.50	2.775	1.570	1.767
0.03	0.0085	11.00	4.070	1.575	2.584
0.04	0.0114	14.00	5.180	1.579	3.280
0.05	0.0142	18.00	6.660	1.584	4.205
0.06	0.0170	21.00	7.770	1.589	4.891
0.10	0.0284	31.00	11.470	1.607	7.137
0.14	0.0398	38.00	14.060	1.626	8.646
0.18	0.0511	44.00	16.280	1.646	9.893
0.22	0.0625	48.00	17.760	1.666	10.663
0.26	0.0739	52.00	19.240	1.686	11.412
0.30	0.0852	53.00	19.610	1.707	11.488
0.35	0.0994	52.00	19.240	1.734	11.097
0.40	0.1136	50.00	18.500	1.762	10.502
0.45	0.1278	49.00	18.130	1.790	10.127
0.50	0.1420	49.00	18.130	1.820	9.962

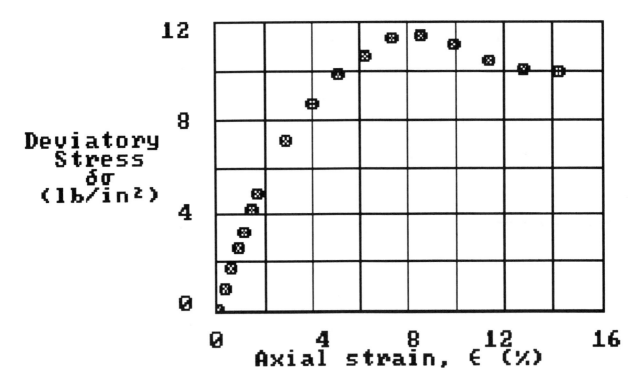

Unconsolidated-Undrained Triaxial Test

CONSOLIDATED-UNDRAINED TRIAXIAL TEST

(Preliminary data)

Description of soil: Dark brown silty clay
Location: Carbondale, Illinois
Specimen No.: 134

Moist unit weight of specimen (beginning of test) (g):	185.65
Moisture content (beginning of test) (%):	35.35
Initial length of specimen, Lo (mm):	76.20
Initial diameter of specimen, Do (mm):	35.70
Initial area of the specimen, Ao (cm²):	10.01
Initial volume of the specimen, Vo (cu cm):	76.27

AFTER CONSOLIDATION OF SATURATED SPECIMEN IN TRIAXIAL CELL

Cell consolidation pressure, σ_3 (kN/m²):	392.00
Net drainage from the specimen during consolidation, δV (cu cm):	11.60
Volume of specimen after consolidation, Vc (cu cm):	64.67
Area of specimen after consolidation, Ac (cm²):	8.97
Length of specimen after consolidation, Lc (in):	2.84

(Axial stress-strain calculation)

Proving ring calibration factor (N/div): 1.0713

Specimen deforma- = δL (in)	Vertical strain, $\epsilon = \dfrac{\delta L}{Lc}$	Proving ring dial reading -No. of small divisions	Piston load = Col. 3 x calibration factor, P (N)	Corrected area $A = \dfrac{Ac}{1-\epsilon}$ (cm²)	Deviatory stress $\delta\sigma = P/A$ (kN/m²)	Excess pore water pressure δu (kN/m²)	$\bar{A} = \dfrac{\delta u}{\delta\sigma}$
0.000	0.0000	0.00	0.000	8.967	0.000	0.00	0.000
0.006	0.0021	15.00	16.070	8.986	17.882	2.94	0.164
0.015	0.0053	109.00	116.772	9.015	129.531	49.05	0.379
0.024	0.0085	135.00	144.626	9.044	159.918	74.56	0.466
0.030	0.0106	147.00	157.481	9.063	173.761	89.27	0.514
0.045	0.0158	172.00	184.264	9.112	202.227	111.83	0.553
0.060	0.0211	192.00	205.690	9.161	224.530	135.38	0.603
0.072	0.0254	205.00	219.617	9.201	238.698	148.13	0.621
0.090	0.0317	225.00	241.043	9.261	260.281	160.88	0.618
0.108	0.0380	236.00	252.827	9.322	271.219	167.75	0.619
0.124	0.0437	247.00	264.611	9.377	282.198	175.60	0.622
0.168	0.0592	265.00	283.895	9.531	297.857	176.58	0.593
0.180	0.0634	270.00	289.251	9.574	302.114	176.58	0.584
0.198	0.0697	278.00	297.821	9.639	308.960	176.58	0.572
0.216	0.0761	284.00	304.249	9.706	313.477	176.58	0.563
0.234	0.0824	287.00	307.463	9.773	314.615	176.58	0.561
0.257	0.0905	288.00	308.534	9.860	312.924	176.57	0.564
0.286	0.1007	286.00	306.392	9.972	307.262	160.88	0.524
0.335	0.1180	275.00	294.608	10.167	289.775	163.82	0.565

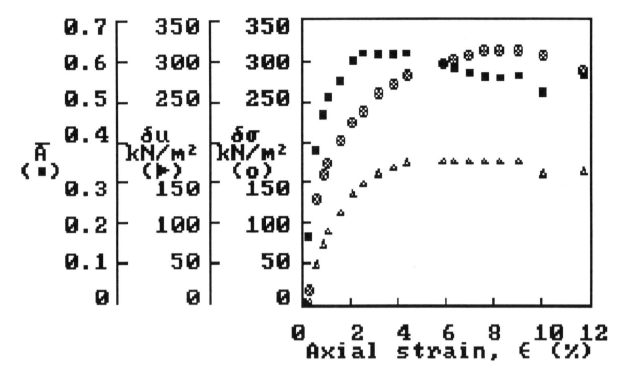

Consolidated-Undrained Triaxial Test

Appendix C
Tables for Laboratory Work

DETERMINATION OF MOISTURE CONTENT

Description of soil_____

Location _____

Sample No. _____

Can No.	Weight of can, W_1 (g)	Weight of can + wet soil, W_2 (g)	Weight of can + dry soil, W_3 (g)	$w(\%) = \dfrac{W_2 - W_3}{W_3 - W_1} \times 100$

DETERMINATION OF MOISTURE CONTENT

Description of soil _____

Location _____

Sample No. _____

Can No.	Weight of can, W_1 (g)	Weight of can + wet soil, W_2 (g)	Weight of can + dry soil, W_3 (g)	$w(\%) = \dfrac{W_2 - W_3}{W_3 - W_1} \times 100$

SPECIFIC GRAVITY

Description of soil _____

Location _____

Sample No. _____

Volume of flask at 20°C _____

Test No.			
Volumetric flask no.			
Weight of flask + water filled to mark, W_1 (g)			
Weight of flask + soil + water filled to mark, W_2 (g)			
Weight of dry soil, W_3 (g)			
Temperature of test, T_1°C			
$G_{s \text{ (at } T_1°C)} = \dfrac{W_3}{(W_1 + W_3) - W_2}$			
A			
$G_{s \text{ (at 20°C)}} = G_{s \text{ (at } T_1°C)} \times A$			

SPECIFIC GRAVITY

Description of soil _____

Location _____

Sample No. _____

Volume of flask at 20°C _____

Test No.			
Volumetric flask no.			
Weight of flask + water filled to mark, W_1 (g)			
Weight of flask + soil + water filled to mark, W_2 (g)			
Weight of dry soil, W_3 (g)			
Temperature of test, $T_1°C$			
$G_{s \, (at \, T_1°C)} = \dfrac{W_3}{(W_1 + W_3) - W_2}$			
A			
$G_{s \, (at \, 20°C)} = G_{s \, (at \, T_1°C)} \times A$			

SIEVE ANALYSIS

Description of soil _____

Location _____

Sample No. _____

Weight of oven dry sample, W _____ (g)

Sieve No.	Sieve opening (mm)	Weight retained on each sieve (g)	Percent of weight retained on each sieve	Cumulative percent retained	Percent finer
Pan	——				

$$\Sigma _____ = W_1$$

Loss during sieve analysis $= \dfrac{W - W_1}{W} \times 100 =$ _____ % (OK if less than 2%)

HYDROMETER ANALYSIS

Description of soil _____

Sample No. _____ Location _____

G_s _____ Hydrometer type _____

Dry weight of soil, W_s _____ (g) Temperature of test, T _____ (°C)

Meniscus correction, F_m _____ Zero correction, F_z _____

Temperature correction, F_T _____

Time (min) (1)	Hydrometer reading, R (2)	R_{cp} (3)	Percent finer, $\dfrac{a\,R_{cp}}{50} \times 100$ (4)	R_{cL} (5)	L (cm) (6)	A (7)	D (mm) (8)

LIQUID LIMIT TEST

Description of soil _____

Location _____

Sample No. _____

Can No.	Weight of can, W_1 (g)	Weight of can + wet soil, W_2 (g)	Weight of can + dry soil, W_3 (g)	Moisture content, w (%)	Number of blows, N

Liquid limit = _____

Flow index = _____

PLASTIC LIMIT TEST

Description of soil _____

Location _____

Sample No. _____

Can No.	Weight of can, W_1 (g)	Weight of can + wet soil, W_2 (g)	Weight of can + dry soil, W_3 (g)	$PL = \dfrac{W_2 - W_3}{W_3 - W_1} \times 100$

Plasticity index $= PI = LL - PL =$ _____ $=$ _____

SHRINKAGE LIMIT TEST

Description of soil _____

Location _____

Sample No. _____

Test No.			
Weight of coated shrinkage limit dish, W_1 (g)			
Weight of dish + wet soil, W_2 (g)			
Weight of dish + dry soil, W_3 (g)			
$w_i = \dfrac{(W_2 - W_3)}{(W_3 - W_1)} \times 100,\ (\%)$			
Weight of mercury to fill the dish, W_4 (g)			
Weight of mercury displaced by soil pat, W_5 (g)			
$\dfrac{(W_4 - W_5)}{(13.6)\ (W_3 - W_1)}\ (100)$			
$SL = w_i - \dfrac{(W_4 - W_5)}{(13.6)\ (W_3 - W_1)}\ (100)$			

SHRINKAGE LIMIT TEST

Description of soil _____

Location _____

Sample No. _____

Test No.			
Weight of coated shrinkage limit dish, W_1 (g)			
Weight of dish + wet soil, W_2 (g)			
Weight of dish + dry soil, W_3 (g)			
$w_i = \dfrac{(W_2 - W_3)}{(W_3 - W_1)} \times 100$, (%)			
Weight of mercury to fill the dish, W_4 (g)			
Weight of mercury displaced by soil pat, W_5 (g)			
$\dfrac{(W_4 - W_5)}{(13.6)\,(W_3 - W_1)}(100)$			
$SL = w_i - \dfrac{(W_4 - W_5)}{(13.6)\,(W_3 - W_1)}(100)$			

CONSTANT HEAD PERMEABILITY TEST

Description of soil _____

Sample No. _____ Location _____

Length of specimen, L_____ (cm) Diameter of specimen, D_____ (cm)

Area of specimen, A_____ (cm^2) Dry weight of specimen, $W_2 - W_1$_____ g

Void ratio, e _____ G_s_____

Test No.	Average flow, Q (cm^3)	Time of collection, t (sec)	Temperature of water, T (°C)	Head difference, h (cm)	$k = \dfrac{QL}{Aht}$ (cm/sec)

Average k = _____ cm/sec

$$k_{(20°C)} = k_{T°C} \frac{\eta_{T°C}}{\eta_{20°C}} = \text{_____} = \text{_____} \text{ cm/sec}$$

FALLING HEAD PERMEABILITY TEST

Description of soil _____

Sample No._____ Location _____

Length of specimen, L_____ (cm) Area of specimen, A_____ (cm^2)

Dry weight of specimen_____ (g) G_s_____ Void ratio, e _____

Test No.	h_1 (cm)	h_2 (cm)	Test duration, t (sec)	Temperature of water, T (°C)	Volume of water, V (cm^3)	$k = \dfrac{2.303\,VL}{(h_1-h_2)\,tA}\log\dfrac{h_1}{h_2}$

$k_{20°C}$ = _____ = _____   Average k = _____ cm/sec

.

FALLING HEAD PERMEABILITY TEST

Description of soil _____

Sample No. _____ Location _____

Length of specimen, L_____ (cm) Area of specimen, A_____ (cm^2)

Dry weight of specimen _____ (g) G_s _____ Void ratio, e _____

Test No.	h_1 (cm)	h_2 (cm)	Test duration, t (sec)	Temperature of water, T (°C)	Volume of water, V (cm^3)	$k = \dfrac{2.303\,VL}{(h_1 - h_2)\,tA} \log \dfrac{h_1}{h_2}$

$k_{20°C} =$ _____ $=$ _____   Average $k =$ _____ cm/sec

PROCTOR COMPACTION TEST

Description of soil _____

Sample No. _____ Location _____

Volume of mold _____ Weight of hammer _____

No. of blows/layer _____ No. of layers _____ G_s _____

Test No.	Weight of mold, W_1 (lb)	Weight of mold + moist soil, W_2 (lb)	Weight of moist soil, $W_2 - W_1$ (lb)	Moist unit weight, $30(W_2 - W_1)$ (lb/ft^3)	Moisture[a] content, w (%)	Dry unit weight, γ_d (lb/ft^3)

[a] To be obtained from the following table.

Moisture Content Determination

Test No.						
Can No.						
Weight of can (g)						
Weight of can + moist soil (g)						
Weight of can + dry soil (g)						
Moisture content (%)						

FIELD UNIT WEIGHT—SAND CONE METHOD

Calibration of Unit Weight of Ottawa Sand

Weight of sand in the mold, W_1 _____

Volume of mold, V_1 _____

$\gamma_{d\,(sand)} = \dfrac{W_1}{V_1}$ _____

Calibration Cone

Weight of bottle + cone + sand (before use), W_2 _____

Weight of bottle + cone + sand (after use), W_3 _____

Weight of sand to fill the cone, $W_c = W_2 - W_3$ _____

Results from Field Tests

Weight of bottle + cone + sand (before use), W_5 _____

Weight of bottle + cone + sand (after use), W_7 _____

Volume of hole, $\dfrac{W_5 - W_7 - W_c}{\gamma_{d\,(sand)}}$ _____

Weight of gallon can, W_4 _____

Weight of can + moist soil, W_6 _____

Weight of can + dry soil, W_8 _____

Weight of moist soil, $W_6 - W_4$ _____

Moist unit weight of soil in field, $\gamma = \dfrac{(W_6 - W_4)\,\gamma_{d\,(sand)}}{W_5 - W_7 - W_c}$ _____

Moisture content in the field, $w\,(\%) = \dfrac{W_6 - W_8}{W_8 - W_4}\,(100)$ _____

Dry unit weight in the field, $\gamma_d = \dfrac{\gamma}{1 + \dfrac{w\%}{100}}$ _____

DIRECT SHEAR TEST ON SAND

Description of soil _____

Sample No. _____ Location _____

Size of specimen _____

Weight of specimen, W_____ Dry unit weight _____

G_s _____ Void ratio, e _____ Normal load, N _____

Normal stress, σ' _____

Proving ring calibration factor _____

Horizontal displacement (in.) (1)	Vertical displacement* (in.) (2)	No. of div. in proving ring dial gauge (3)	Shear force, S (lb) (4)	Shear stress, τ (lb/in.²) (5)

* + sign means expansion

DIRECT SHEAR TEST ON SAND

Description of soil _____

Sample No. _____ Location _____

Size of specimen _____

Weight of specimen, W_____ Dry unit weight _____

G_s _____ Void ratio, e _____ Normal load, N _____

Normal stress, σ' _____

Proving ring calibration factor _____

Horizontal displacement (in.) (1)	Vertical displacement* (in.) (2)	No. of div. in proving ring dial gauge (3)	Shear force, S (lb) (4)	Shear stress, τ (lb/in.²) (5)

* + sign means expansion

DIRECT SHEAR TEST ON SAND

Description of soil _____

Sample No. _____ Location _____

Size of specimen _____

Weight of specimen, W_____ Dry unit weight _____

G_s _____ Void ratio, e _____ Normal load, N _____

Normal stress, σ' _____

Proving ring calibration factor _____

Horizontal displacement (in.) (1)	Vertical displacement* (in.) (2)	No. of div. in proving ring dial gauge (3)	Shear force, S (lb) (4)	Shear stress, τ (lb/in.²) (5)

* + sign means expansion

DIRECT SHEAR TEST ON SAND

Description of soil _____

Sample No. _____ Location _____

Size of specimen _____

Weight of specimen, W_____ Dry unit weight _____

G_s _____ Void ratio, e _____ Normal load, N _____

Normal stress, σ' _____

Proving ring calibration factor _____

Horizontal displacement (in.) (1)	Vertical displacement* (in.) (2)	No. of div. in proving ring dial gauge (3)	Shear force, S (lb) (4)	Shear stress, τ (lb/in.²) (5)

* + sign means expansion

DIRECT SHEAR TEST ON SAND

Description of soil _____

Sample No. _____ Location _____

Size of specimen _____

Weight of specimen, W _____ Dry unit weight _____

G_s _____ Void ratio, e _____ Normal load, N _____

Normal stress, σ' _____

Proving ring calibration factor _____

Horizontal displacement (in.) (1)	Vertical displacement* (in.) (2)	No. of div. in proving ring dial gauge (3)	Shear force, S (lb) (4)	Shear stress, τ (lb/in.^{2}) (5)

* + sign means expansion

DIRECT SHEAR TEST ON SAND

Description of soil _____

Sample No. _____ Location _____

Size of specimen _____

Weight of specimen, W_____ Dry unit weight _____

G_s _____ Void ratio, e _____ Normal load, N _____

Normal stress, σ' _____

Proving ring calibration factor _____

Horizontal displacement (in.) (1)	Vertical displacement* (in.) (2)	No. of div. in proving ring dial gauge (3)	Shear force, S (lb) (4)	Shear stress, τ (lb/in.^{2}) (5)

* + sign means expansion

UNCONFINED COMPRESSION TEST

Description of soil_____

Sample No._____ Location_____

Moist weight of specimen _____ Moisture content_____%

Length of specimen, L_____ Diameter, D_____ Area, A_0_____

Proving ring calibration factor: 1 div. = _____

Specimen deformation = ΔL (in.) (1)	Vertical strain, $\epsilon = \dfrac{\Delta L}{L}$ (2)	Proving ring dial reading [No. of small divisions] (3)	Load = Col. 3 × calibration factor of proving ring (lb) (4)	Corrected area = $A_c = \dfrac{A_0}{1-\epsilon}$ (in.²) (5)	Stress = $\dfrac{\text{Col. 4}}{\text{Col. 5}}$ (lb/in.²) (6)

CONSOLIDATION TEST
(Time vs. vertical dial reading)

Description of soil _____

Pressure on specimen _____

Clock time of load application _____

Time after load application, t (min.)	$\sqrt{t}$ (min$^{0.5}$)	Vertical dial reading (in.)

CONSOLIDATION TEST
(Time vs. vertical dial reading)

Description of soil _____ _____

Pressure on specimen _____ _____

Clock time of load application _____ _____

Time after load application, t (min.)	$\sqrt{t}$ (min$^{0.5}$)	Vertical dial reading (in.)

CONSOLIDATION TEST
(Time vs. vertical dial reading)

Description of soil _____

Pressure on specimen _____

Clock time of load application _____

Time after load application, t (min.)	$\sqrt{t}$ (min$^{0.5}$)	Vertical dial reading (in.)

CONSOLIDATION TEST
(Time vs. vertical dial reading)

Description of soil _____

Pressure on specimen _____

Clock time of load application _____

Time after load application, t (min.)	$\sqrt{t}$ (min$^{0.5}$)	Vertical dial reading (in.)

CONSOLIDATION TEST
(Time vs. vertical dial reading)

Description of soil _____

Pressure on specimen _____

Clock time of load application _____

Time after load application, t (min.)	$\sqrt{t}$ (min$^{0.5}$)	Vertical dial reading (in.)

CONSOLIDATION TEST
(Time vs. vertical dial reading)

Description of soil _____

Pressure on specimen _____

Clock time of load application _____

Time after load application, t (min.)	$\sqrt{t}$ (min$^{0.5}$)	Vertical dial reading (in.)

CONSOLIDATION TEST
(Time vs. vertical dial reading)

Description of soil _____

Pressure on specimen _____

Clock time of load application _____

Time after load application, t (min.)	$\sqrt{t}$ (min$^{0.5}$)	Vertical dial reading (in.)

CONSOLIDATION TEST

(Void ratio-pressure and coefficient of consolidation calculation)

Description of soil _Patapsco Clay, bluish-green color_   Location _Road Embankment IC95, Baltimore_

Specimen diameter _2.5_   Initial specimen height, $H_{t(i)}$ _1_   inch = _.2067_ reading

Moisture content: Beginning of test _35.8_   (%) End of test _____ %

Weight of dry soil specimen _____ G_s _____   Height of solids, H_s _____   cm = _____

Pressure, p (ton/ft²)	Final dial reading (in.)	Change in specimen height (in.)	Final specimen height, $H_{t(f)}$ (in.)	Height of void, H_v (in.)	Final void ratio, e	Average height during consolidation, $H_{t(av)}$ (in.)	Fitting time (sec)		c_v from × 10³ (in.²/sec)	
							t_{90}	t_{50}	t_{90}	t_{50}
(1)	(2)	(3)	(4)	(5)	(6)	(7)	(8)	(9)	(10)	(11)
.25	.1784	.0283								
.5	.1650	.0417								
1	.1440	.0607								
2	.1055	.1012								
4	.0011	.2056								
8	.0008	.2059								

UNCONSOLIDATED-UNDRAINED TRIAXIAL TEST
(Preliminary data)

Description of soil _____

Location _____

Specimen No. _____

Moist unit weight of specimen (end of test)   _____

Dry unit weight of specimen   _____

Moisture content (end of test)   _____

Initial average length of specimen, L_0 _____

Initial average diameter of specimen, D_0 _____

Initial area, A_0 _____

G_s _____

Final degree of saturation   _____

Cell confining pressure, σ_3 _____

Proving ring calibration factor   _____

UNCONSOLIDATED-UNDRAINED TRIAXIAL TEST
(Preliminary data)

Description of soil _____

Location _____

Specimen No. _____

Moist unit weight of specimen (end of test) _____

Dry unit weight of specimen _____

Moisture content (end of test) _____

Initial average length of specimen, L_0 _____

Initial average diameter of specimen, D_0 _____

Initial area, A_0 _____

G_s _____

Final degree of saturation _____

Cell confining pressure, σ_3 _____

Proving ring calibration factor _____

UNCONSOLIDATED-UNDRAINED TRIAXIAL TEST
(Preliminary data)

Description of soil _____

Location _____

Specimen No. _____

Moist unit weight of specimen (end of test)   _____

Dry unit weight of specimen   _____

Moisture content (end of test)   _____

Initial average length of specimen, L_0 _____

Initial average diameter of specimen, D_0 _____

Initial area, A_0 _____

G_s _____

Final degree of saturation   _____

Cell confining pressure, σ_3 _____

Proving ring calibration factor   _____

UNCONSOLIDATED-UNDRAINED TRIAXIAL TEST
(Axial stress-strain calculation)

Specimen deformation $= \Delta L$ (in.) (1)	Vertical strain, $\epsilon = \dfrac{\Delta L}{L_0}$ (2)	Proving ring dial reading [No. of small divisions] (3)	Piston load, P [Col. 3 $\times$ calibration factor] (lb) (4)	Corrected area, $A = \dfrac{A_0}{1-\epsilon}$ (in.^{2}) (5)	Deviatory stress, $\Delta \sigma = \dfrac{P}{A}$ (lb/in.^{2}) (6)

UNCONSOLIDATED-UNDRAINED TRIAXIAL TEST
(Axial stress-strain calculation)

Specimen deformation $= \Delta L$ (in.) (1)	Vertical strain, $\epsilon = \dfrac{\Delta L}{L_0}$ (2)	Proving ring dial reading [No. of small divisions] (3)	Piston load, P [Col. 3 $\times$ calibration factor] (lb) (4)	Corrected area, $A = \dfrac{A_0}{1 - \epsilon}$ (in.^{2}) (5)	Deviatory stress, $\Delta\sigma = \dfrac{P}{A}$ (lb/in.^{2}) (6)

UNCONSOLIDATED-UNDRAINED TRIAXIAL TEST
(Axial stress-strain calculation)

Specimen deformation $= \Delta L$ (in.) (1)	Vertical strain, $\epsilon = \dfrac{\Delta L}{L_0}$ (2)	Proving ring dial reading [No. of small divisions] (3)	Piston load, P [Col. 3 $\times$ calibration factor] (lb) (4)	Corrected area, $A = \dfrac{A_0}{1 - \epsilon}$ (in.^{2}) (5)	Deviatory stress, $\Delta \sigma = \dfrac{P}{A}$ (lb/in.^{2}) (6)

UNCONSOLIDATED-UNDRAINED TRIAXIAL TEST
(Axial stress-strain calculation)

Specimen deformation $= \Delta L$ (in.) (1)	Vertical strain, $\epsilon = \dfrac{\Delta L}{L_0}$ (2)	Proving ring dial reading [No. of small divisions] (3)	Piston load, P [Col. 3 $\times$ calibration factor] (lb) (4)	Corrected area, $A = \dfrac{A_0}{1-\epsilon}$ (in.^{2}) (5)	Deviatory stress, $\Delta \sigma = \dfrac{P}{A}$ (lb/in.^{2}) (6)

UNCONSOLIDATED-UNDRAINED TRIAXIAL TEST
(Axial stress-strain calculation)

Specimen deformation $= \Delta L$ (in.) (1)	Vertical strain, $\epsilon = \dfrac{\Delta L}{L_0}$ (2)	Proving ring dial reading [No. of small divisions] (3)	Piston load, P [Col. 3 × calibration factor] (lb) (4)	Corrected area, $A = \dfrac{A_0}{1 - \epsilon}$ (in.^{2}) (5)	Deviatory stress, $\Delta\sigma = \dfrac{P}{A}$ (lb/in.^{2}) (6)

UNCONSOLIDATED-UNDRAINED TRIAXIAL TEST
(Axial stress-strain calculation)

Specimen deformation $= \Delta L$ (in.) (1)	Vertical strain, $\epsilon = \dfrac{\Delta L}{L_0}$ (2)	Proving ring dial reading [No. of small divisions] (3)	Piston load, P [Col. 3 $\times$ calibration factor] (lb) (4)	Corrected area, $A = \dfrac{A_0}{1 - \epsilon}$ (in.^{2}) (5)	Deviatory stress, $\Delta\sigma = \dfrac{P}{A}$ (lb/in.^{2}) (6)

CONSOLIDATED-UNDRAINED TRIAXIAL TEST
(Preliminary data)

Description of soil _____

Location _____

Specimen No. _____

Moist unit weight of specimen (beginning of test)   _____

Moisture content (beginning of test)   _____

Initial length of specimen, L_0 _____

Initial diameter of specimen, D_0 _____

Initial area of the specimen, $A_0 = \dfrac{\pi}{4}D_0^2$

_____

Initial volume of the specimen, $V_0 = A_0 L_0$ _____

After Consolidation of Saturated Specimen in Triaxial Cell

Cell consolidation pressure, σ_3 _____

Net drainage from the specimen during consolidation, ΔV

Volume of specimen after consolidation, $V_0 - \Delta V = V_c$ _____

Area of specimen after consolidation,

$$A_c = A_0 \left(\frac{V_c}{V_0} \right)^{2/3}$$

_____

Length of the specimen after consolidation,

$$L_c = L_0 \left(\frac{V_c}{V_0} \right)^{1/3}$$

_____

CONSOLIDATED-UNDRAINED TRIAXIAL TEST
(Preliminary data)

Description of soil _____

Location _____

Specimen No. _____

Moist unit weight of specimen (beginning of test)   _____

Moisture content (beginning of test)   _____

Initial length of specimen, L_0 _____

Initial diameter of specimen, D_0 _____

Initial area of the specimen, $A_0 = \dfrac{\pi}{4}D_0^2$

_____

Initial volume of the specimen, $V_0 = A_0L_0$ _____

After Consolidation of Saturated Specimen in Triaxial Cell

Cell consolidation pressure, σ_3 _____

Net drainage from the specimen during
consolidation, ΔV _____

Volume of specimen after consolidation,
$V_0 - \Delta V = V_c$ _____

Area of specimen after consolidation,

$$A_c = A_0\left(\frac{V_c}{V_0}\right)^{2/3}$$

_____

Length of the specimen after consolidation,

$$L_c = L_0\left(\frac{V_c}{V_0}\right)^{1/3}$$

_____

CONSOLIDATED-UNDRAINED TRIAXIAL TEST
(Preliminary data)

Description of soil _____

Location _____

Specimen No. _____

Moist unit weight of specimen (beginning of test) _____

Moisture content (beginning of test) _____

Initial length of specimen, L_0 _____

Initial diameter of specimen, D_0 _____

Initial area of the specimen, $A_0 = \dfrac{\pi}{4} D_0^2$ _____

Initial volume of the specimen, $V_0 = A_0 L_0$ _____

After Consolidation of Saturated Specimen in Triaxial Cell

Cell consolidation pressure, σ_3 _____

Net drainage from the specimen during consolidation, ΔV _____

Volume of specimen after consolidation, $V_0 - \Delta V = V_c$ _____

Area of specimen after consolidation,

$$A_c = A_0 \left(\frac{V_c}{V_0} \right)^{2/3}$$

_____

Length of the specimen after consolidation,

$$L_c = L_0 \left(\frac{V_c}{V_0} \right)^{1/3}$$

_____

CONSOLIDATED-UNDRAINED TRIAXIAL TEST
(Preliminary data)

Description of soil _____

Location _____

Specimen No. _____

Moist unit weight of specimen (beginning of test)   _____

Moisture content (beginning of test)   _____

Initial length of specimen, L_0 _____

Initial diameter of specimen, D_0 _____

Initial area of the specimen, $A_0 = \frac{\pi}{4}D_0^2$ _____

Initial volume of the specimen, $V_0 = A_0L_0$ _____

After Consolidation of Saturated Specimen in Triaxial Cell

Cell consolidation pressure, σ_3 _____

Net drainage from the specimen during consolidation, ΔV _____

Volume of specimen after consolidation, $V_0 - \Delta V = V_c$ _____

Area of specimen after consolidation,

$$A_c = A_0\left(\frac{V_c}{V_0}\right)^{2/3}$$

_____

Length of the specimen after consolidation,

$$L_c = L_0\left(\frac{V_c}{V_0}\right)^{1/3}$$

_____

CONSOLIDATED-UNDRAINED TRIAXIAL TEST
(Preliminary data)

Description of soil _____

Location _____

Specimen No. _____

Moist unit weight of specimen (beginning of test)　　_____

Moisture content (beginning of test)　　_____

Initial length of specimen, L_0　　_____

Initial diameter of specimen, D_0　　_____

Initial area of the specimen,　$A_0 = \dfrac{\pi}{4}D_0^2$

_____

Initial volume of the specimen, $V_0 = A_0 L_0$　　_____

After Consolidation of Saturated Specimen in Triaxial Cell

Cell consolidation pressure, σ_3　　_____

Net drainage from the specimen during consolidation, ΔV　　_____

Volume of specimen after consolidation, $V_0 - \Delta V = V_c$　　_____

Area of specimen after consolidation,

$$A_c = A_0\left(\frac{V_c}{V_0}\right)^{2/3}$$

_____

Length of the specimen after consolidation,

$$L_c = L_0\left(\frac{V_c}{V_0}\right)^{1/3}$$

_____

~~.432~~	~~377~~	~~56.7~~
.745 .962	384	58.3
~~.942~~	~~388~~	~~56~~
.807 1.024	392	55.8
~~1.052~~	~~396~~	~~55.5~~
.867 1.044	400	~~55~~
~~1.113~~	~~404~~	~~54.6~~
.928 1.145	408	54.3
~~1.175~~	~~411~~	~~53.9~~
.988 1.205	415	53.6
~~1.236~~	~~418~~	~~53.3~~
1.049 1.266	421	52.4
~~1.296~~	~~425~~	~~52.8~~
1.112 1.329	428	52.3
~~1.177~~ ~~1.359~~	~~431~~	~~52.1~~
1.112 1.389	434	51.7

5.u in

CONSOLIDATED-UNDRAINED TRIAXIAL TEST
(Axial stress-strain calculation)

Proving ring calibration factor_____

Specimen deformation $= \Delta L$ (in.) (1)	Vertical strain, $\epsilon = \dfrac{\Delta L}{L}$ (2)	Proving ring dial reading [No. of small divisions] (3)	Piston load, D [Col. 3 × calibration factor] (N) (4)	Corrected area $= A = \dfrac{A_c}{1-\epsilon}$ (cm^2) (5)	Diviatory stress $\Delta\sigma = \dfrac{P}{A}$ (kN/m^2) (6)	Excess pore water pressure, Δu (kN/m^2) (7)	$\bar{A} = \dfrac{\Delta L}{\Delta\sigma}$ (8)
.217			11			(41.5)	
.628 .245			121			44.5	
~~.265~~			~~154~~			~~46.1~~	
.076 .293			179			48.2	
~~.322~~			~~197~~			~~50.3~~	
.135 .352			213			52.2	
~~.384~~			~~226~~			~~53.7~~	
.196 .413			234			55	
~~.443~~			~~250~~			~~56~~	
.258 .474			261			56.8	
~~.504~~			~~272~~			~~57.2~~	
.318 .535			282	282		58.4	
~~.565~~			~~291~~			~~58.3~~	
.378 .595			301			58.5	
~~.625~~			~~310~~			~~58.6~~	
.440 .657			319			58.7	
~~.687~~			~~327~~			~~58.5~~	
.50 .718			335			58.6	
~~.748~~			~~343~~			~~58.4~~	
.56 .779			350			58.2	
~~.810~~			357			~~58.0~~	
.625 .842			363			57.7	
~~.864~~			~~364~~			~~57.4~~	
.685 .902			374			57	

CONSOLIDATED-UNDRAINED TRIAXIAL TEST
(Axial stress-strain calculation)

Proving ring calibration factor_____

Specimen deformation $= \Delta L$ (in.) (1)	Vertical strain, $\epsilon = \dfrac{\Delta L}{L}$ (2)	Proving ring dial reading [No. of small divisions] (3)	Piston load, D [Col. 3 $\times$ calibration factor] (N) (4)	Corrected area = $A = \dfrac{A_c}{1-\epsilon}$ (cm^2) (5)	Diviatory stress $\Delta\sigma = \dfrac{P}{A}$ (kN/m^2) (6)	Excess pore water pressure, Δu (kN/m^2) (7)	$\bar{A} = \dfrac{\Delta L}{\Delta\sigma}$ (8)
		110	33			3	
		168	43			6.7	
		202	45			10.7	
		228	52			13.3	
		250	57			15.3	
		271	62			15.9	
		296	67			17	
		308	72			17.2	
		324	76			17.1	
		339	82			16.7	
		352	86		50.72	16.2	
		363	90		51.70	15.5	
		373	95		52.53	14.8	
		381	99		52.91		
		389	103		53.36	13.5	
		397	108		53.79	12.8	
		404	111		54.01	12.1	
		416	116		54.08	10.9	
		417	119		54.22	10.8	
		423	123		54.3	10.2	

126
130
133

CONSOLIDATED-UNDRAINED TRIAXIAL TEST
(Axial stress-strain calculation)

P92

Proving ring calibration factor_____

Specimen deformation $= \Delta L$ (in.) (1)	Vertical strain, $\epsilon = \dfrac{\Delta L}{L}$ (2)	Proving ring dial reading [No. of small divisions] (3)	Piston load, D [Col. 3 × calibration factor] (N) (4)	Corrected area = $A = \dfrac{A_c}{1-\epsilon}$ (cm^2) (5)	Diviatory stress $\Delta\sigma = \dfrac{P}{A}$ (kN/m^2) (6)	Excess pore water pressure, Δu (kN/m^2) (7)	$\bar{A} = \dfrac{\Delta L}{\Delta\sigma}$ (8)
			136				
			138				
			140				
			143				
			144				
			147				
			150				
			151				
			153				
			154				
			100				
			102				
			104				
			166				

CONSOLIDATED-UNDRAINED TRIAXIAL TEST
(Axial stress-strain calculation)

Proving ring calibration factor_____

Specimen deformation $= \Delta L$ (in.) (1)	Vertical strain, $\epsilon = \dfrac{\Delta L}{L}$ (2)	Proving ring dial reading [No. of small divisions] (3)	Piston load, D [Col. 3 $\times$ calibration factor] (N) (4)	Corrected area $= A = \dfrac{A_c}{1 - \epsilon}$ (cm^2) (5)	Diviatory stress $\Delta\sigma = \dfrac{P}{A}$ (kN/m^2) (6)	Excess pore water pressure, Δu (kN/m^2) (7)	$\bar{A} = \dfrac{\Delta L}{\Delta\sigma}$ (8)

CONSOLIDATED-UNDRAINED TRIAXIAL TEST
(Axial stress-strain calculation)

Proving ring calibration factor_____

Specimen deformation $= \Delta L$ (in.) (1)	Vertical strain, $\epsilon = \dfrac{\Delta L}{L}$ (2)	Proving ring dial reading [No. of small divisions] (3)	Piston load, D [Col. 3 × calibration factor] (N) (4)	Corrected area = $A = \dfrac{A_c}{1-\epsilon}$ (cm^2) (5)	Diviatory stress $\Delta\sigma = \dfrac{P}{A}$ (kN/m^2) (6)	Excess pore water pressure, Δu (kN/m^2) (7)	$\bar{A} = \dfrac{\Delta L}{\Delta\sigma}$ (8)